AF865425

COLLEGE MATH PLACEMENT TEST PREP FOR ADULT LEARNERS

A 30-Day Refresher for Returning Students

MATTHEW TOM BURROWS

Rebuild the Math That Matters

This book provides general educational information and practice. It is not an official publication of any college, university, state agency, testing company, or professional licensing body.

No score, placement, admission, academic, financial, or career result is guaranteed. Readers are responsible for confirming current test formats, calculator rules, deadlines, retest policies, course requirements, and placement decisions directly with the relevant institution.

This book is not a substitute for individual advice from a college adviser, instructor, tutor, accessibility specialist, or other qualified professional. Names and scenarios are composites or are used in an appropriate educational manner. They are not intended to identify, harm, or misrepresent any person or organization. ACCUPLACER is a registered trademark of College Board. ALEKS is a registered trademark of ALEKS Corporation. All trademarks are used only for descriptive purposes, and no endorsement is implied.

ISBN: 978-1-911838-31-9

Table Of Contents

Preface

Math placement is not one national exam with one passing score. The college, program, Target Course, and local Placement Policy determine what a result means. For a returning student, the first task is therefore not to review every topic once taught in school. It is to identify the exact route ahead, then rebuild the mathematics that route requires.

The central method in this book is the Five-Rung Placement Ladder: Clarify, Diagnose, Rebuild, Rehearse, and Decide. It is not a general survey of every mathematics topic taught in school, and it is not a promise to force the highest possible score. It is a structured way to identify the mathematics your institution and Target Course actually require, separate Ready Skills from Recall Skills and Rebuild Skills, repair the most useful gaps, practice in the correct format, and make a sound decision after the result arrives.

The gap this book addresses is practical. Returning students can usually find large test-preparation guides, broad adult math refreshers, free online lessons, and official sample questions. What is harder to find is a single plan that connects those resources to an adult schedule and a real college requirement. Without that connection, people often study the wrong assessment, spend too much time on low-priority topics, or mistake one difficult practice session for proof that they cannot succeed.

This handbook is built to prevent that waste. It begins with the institution, the placement policy, and the Target Course. It then uses the No-Shame Diagnostic to sort skills into three useful groups. From there, the 30-day plan offers Foundation, College-Ready, and Higher-Math routes, with 15-minute, 30-minute, and 45-minute options. The mathematics sequence moves from number sense through fractions, percentages, ratios, algebra, graphs, data, word problems, and optional higher math. Two practice assessments, worked explanations, an Error Pattern Log, and the Calm Test Routine turn review into a repeatable process.

The method draws on official test descriptions, current placement guidance, verified learning research, and independently checked mathematics examples. It is written for adults returning to community college, university, nursing prerequisites, technical programs, and degree-completion study. Tutors, advisers, instructors, and family members may also use the worksheets to support a returning student, but the primary voice speaks directly to the person preparing for placement.

This is a stand-alone handbook. It is not an official publication of a college or testing company, and it cannot guarantee a score, course placement, admission decision, or program outcome. Policies can change, and local rules take priority over any general explanation in these pages. Confirm current requirements with your institution, and seek help from an adviser, instructor, tutor, accessibility office, or other appropriate professional when your situation requires it.

Its purpose is simple: help you prepare for the mathematics that matters, earn the most accurate placement you can, and begin your next course with a stronger foundation.

How to Use This Book

This is a handbook, not a book you must read straight through before touching a pencil.

Start with the Introduction and Chapters 1 through 5. Those sections help you identify your exact Placement Test, your Placement Policy, your Target Course, and your current skill level. Then follow the study route assigned by the No-Shame Diagnostic.

Use the same notebook throughout the 30 days. Divide it into four sections:

1. Worked examples
2. Error Pattern Log
3. Formula and vocabulary notes
4. Questions for your college or adviser

Do the problems before reading the answers. A worked solution feels obvious when it is already on the page. The test will ask you to begin with a blank space.

The goal is not to force yourself into the highest mathematics course offered. The goal is to earn the strongest placement that matches both your program and your readiness.

Introduction: You Are Rusty, Not Incapable

Maria sat at the kitchen table with a college portal open on her laptop and a yellow legal pad beside it. She had already completed the application, requested an old transcript, and rearranged two work shifts so she could attend orientation. Nursing school had been a private idea for years. Now it had dates attached to it.

Then she clicked a link labeled "Mathematics Placement."

The sample page showed fractions, negative numbers, and an equation with a letter on both sides. Maria recognized every symbol. She could not remember what to do with any of them.

She closed the page.

For the next three days she did everything except return to it. She compared tuition plans. She checked bus times. She ordered a new lunch container. Each task looked sensible. None of them moved her closer to the test.

On the fourth evening, she printed five sample questions and covered the answer key with a sheet of paper. The first question asked for 25 percent of 80. Maria wrote 20. The second asked her to add two fractions. She stalled. Then she drew two rectangles, divided them into equal pieces, and found the answer.

Something important changed in that small moment. She had not recovered all of school mathematics. She had proved that the knowledge was not gone in one solid block. Some of it was ready. Some of it needed a prompt. Some of it needed to be rebuilt.

Maria is a composite example. Her situation combines details common to adults returning to college after years of work, family responsibility, military service, caregiving, or interrupted education.

Here is the principle that will guide the entire book:

> ***You are not preparing by measuring how much mathematics you have forgotten. You are preparing by finding the shortest honest route from your present skills to your Target Course.***

That route is the Five-Rung Placement Ladder:

1. Clarify
2. Diagnose
3. Rebuild
4. Rehearse
5. Decide

You will use those five rungs throughout the next 30 days.

You Are Not the Exception

Adults returning to education often feel as though everyone else arrives with recent classroom knowledge. That picture is false. In fall 2023, more than 6.26 million students age 25 and older were enrolled in United States postsecondary institutions (National Center for Education Statistics, n.d.).

That number does not prove that every older student takes a mathematics Placement Test. It does show that adult learners are not a small curiosity at the edge of higher education. They are part of the ordinary student population.

Age also does not tell you exactly what support a person needs. A 42 year old accountant may be strong with percentages and weak with algebraic notation. A 27 year old returning after military service may remember equations but make mistakes with fractions. A 53 year old health care worker may use ratios every day without calling them ratios.

The No-Shame Diagnostic in this book treats that unevenness as useful information.

Three Kinds of Skill

Every topic you meet belongs in one of three categories.

A **Ready Skill** is something you can do accurately without immediate review. You may feel slow, but the method is available.

A **Recall Skill** is something you learned before and can recover after a brief explanation or example.

A **Rebuild Skill** needs direct instruction and repeated practice. Perhaps the topic was never taught clearly. Perhaps it never settled into long term memory. Perhaps several smaller gaps are blocking it.

These labels are not grades. They are directions.

If a skill is Ready, maintain it with occasional Mixed Practice. If it is Recall, review it and return to it after a day or two. If it is Rebuild, slow down, work through examples, and ask for help when the same confusion continues.

A diagnostic is a map, not a verdict.

Highest Score Is Not Always Best

It is tempting to make "place as high as possible" the only goal. That promise sells test preparation books because it sounds simple. Real college planning is less simple.

Suppose your Target Course is introductory statistics. Spending half your month on advanced polynomial factoring may not improve your route. Suppose you qualify narrowly for calculus but cannot reliably rearrange a formula or work with fractions. Starting one course lower, or taking a supported version of the course, may protect your progress.

The useful question is not, "How high can I force the score?"

It is, "What starting point gives me a fair chance to complete the mathematics my program requires?"

College Board states that institutions set their own ACCUPLACER placement scores and policies. ACCUPLACER itself does not impose a universal course score (College Board, n.d.-i). Some institutions also use multiple factors such as previous grades or coursework, rather than one test result alone (College Board, n.d.-h).

That is why Clarify comes before Rebuild.

The 30-Day Promise

Thirty days can be enough time to make meaningful progress when the task is focused. It is not enough time to master every topic from arithmetic through precalculus.

This plan works by narrowing the task.

You will:

verify the test and policy used by your institution
identify the course you actually need
complete the No-Shame Diagnostic
choose a Foundation, College-Ready, or Higher-Math route
rebuild High-Value Skills
use an Error Pattern Log
complete Mixed Practice
rehearse the Calm Test Routine
interpret your result before making the next decision

The plan offers 15 minute, 30 minute, and 45 minute versions. More time can help, but consistency matters more than one heroic weekend followed by six empty days.

Research on mathematics learning supports spreading practice across time rather than concentrating all practice into one block. In a college calculus study, spaced retrieval practice improved end of semester retention even though performance was lower on the first two spaced practice questions than on comparable massed questions (Lyle et al., 2022). The practical translation is plain: return to a skill after some forgetting has begun.

You do not need to enjoy that slight struggle. You need to recognize what it is doing.

Your First Action

Open your notebook and write four headings:

My Target Program

My Placement Deadline

What I Think I Remember

What I Fear the Test Will Prove

Write for five minutes.

The last heading matters because fear often hides inside practical language. "I need to review algebra" may also mean "I am afraid that one bad score will prove I waited too long." Both statements deserve an answer, but they are not the same problem.

The mathematics chapters will help with algebra. The Five-Rung Placement Ladder will keep the score in its proper place.

Principles to Remember

You are not starting from nothing. You are starting from a mixture of Ready Skills, Recall Skills, and Rebuild Skills.

The Placement Test is not a universal judgement. It is one piece of an institution's Placement Policy. Your first job is to Clarify what your school uses and what your Target Course requires.

Thirty days will not turn every gap into mastery. It can turn confusion into a plan, weak foundations into working skills, and panic into a repeatable process.

Do not climb every mathematics staircase.

Climb the five rungs between where you are and the course you actually need.

Chapter 1: What Placement Really Decides

Aisha had been accepted into a community college health sciences program when an email told her to complete "placement requirements." She read the phrase as a threat.

By lunchtime she had built an entire future around a test she had not taken. A low score would mean remedial mathematics. Remedial mathematics would mean another semester. Another semester would mean more tuition. More tuition would mean she would have to postpone reducing her work hours. By the end of the day, a 45-minute assessment had somehow become evidence that college was a mistake.

The next morning, Aisha called the testing office.

The staff member asked three questions. Had she completed college level mathematics before? How old was her transcript? Which course did her program require?

Aisha learned that the college used more than one factor for placement. Her previous coursework might count. The test did not affect admission. A supported version of the required course was available. Retesting was possible after review.

Nothing about Aisha's mathematics changed during that phone call. The meaning of the test changed.

Aisha is a composite example. The details represent common placement questions, but every institution can answer them differently.

The principle of this chapter is simple:

> ***A Placement Test can influence your starting course, but the Placement Policy determines what the score actually means. Never let an unknown policy become a private catastrophe.***

Placement Is Not Admission

A placement assessment usually helps a school decide which course or level of support fits a student's current preparation. It is not automatically an entrance examination, and it is not automatically pass or fail.

College Board describes ACCUPLACER as a suite intended to support course placement. It also states that each institution establishes its own placement practices and score requirements (College Board, n.d.-g; College Board, n.d.-b).

Florida's Department of Education describes PERT as an untimed computer adaptive assessment used to determine readiness for college level courses. It explicitly says PERT is not a pass or fail test (Florida Department of Education, n.d.).

College Board describes TSIA2 as an assessment of readiness for college level work for nonexempt students. The result may be considered alongside academic background, goals, and interests (College Board, 2020).

Those descriptions share a pattern. Placement helps a school recommend a starting point. The exact consequence comes from the institution or state system using the result.

That distinction gives you your first practical rule:

> ***Do not study a consequence you have only imagined. Verify it.***

Four Possible Placement Outcomes

Your school may use different names, but mathematics placement often leads toward one of four broad outcomes.

The first is direct placement into the required college level course. That might be quantitative reasoning, statistics, college algebra, technical mathematics, or another course.

The second is a college level course with added support. This is often called a corequisite model. You take the credit bearing course and receive extra instruction, lab time, tutoring, or a companion support course at the same time.

The third is a prerequisite foundation course. You complete preparatory mathematics before enrolling in the required college level course.

The fourth is an alternative placement route. A college may use previous grades, completed coursework, standardized test scores, guided self placement, an adviser review, or a combination of measures.

You should not assume that all four exist at your school. You should know the vocabulary so you can ask.

Carlos, 36, searched his college website for "remedial math" and found nothing. He assumed the college had removed support. An adviser explained that the school now used the phrase "corequisite support." Carlos had been searching for an older label, not an absent option.

Words change. The decision still matters.

What Developmental Education Means

A developmental mathematics course is designed to build skills needed before or alongside college level work. Because developmental courses may not count toward a degree even when students pay tuition or use financial aid for them, they can affect cost and program timing (Community College Research Center, 2021).

That history is one reason placement creates strong emotions. Traditional prerequisite sequences have often required students to complete one or more developmental courses that do not count toward the degree before reaching a gateway course. Research organizations have documented reforms intended to reduce unnecessary barriers, including multiple measures placement and corequisite support (Community College Research Center, 2021).

Corequisite support is not simply a nicer name for the same sequence. In a corequisite model, the student usually enrolls in the college level course while receiving additional support. Research has found improved gateway course completion in some corequisite implementations compared with traditional prerequisite remediation, although the evidence is strongest for specific systems, course designs, and student groups (Ran, 2020).

This book will not tell you that every foundation course is harmful. It will not tell you that every higher placement is wise.

A course can be an unnecessary barrier for one student and a useful bridge for another.

Jordan, 29, placed one level below the course required for a construction management program. He wanted to appeal immediately. When he reviewed the diagnostic report, he saw that his arithmetic was strong but his equation skills were fragile. He chose a shorter supported route offered by the college. That decision added weekly lab time but did not add a semester.

Renee, 48, placed below college algebra after rushing through a test she did not know she could retake. Her transcript showed recent success in a college statistics course. An adviser reviewed the record and changed the placement.

The score did not contain the whole story in either case.

Multiple Measures

A Multiple Measures Placement system uses more than one source of evidence. Depending on local policy, measures may include:

high school grade point average

grades in previous mathematics courses

prior college or transfer coursework

other test scores

institution-specific assessments

nonacademic factors such as study habits or learning strategies

College Board recommends using academic and nonacademic information alongside ACCUPLACER scores. A randomized study in SUNY community colleges found that students moved into college-level courses through one multiple-measures model were more likely to enroll in and complete college-level mathematics or English, although the authors emphasized that model design matters (College Board, n.d.-h; Kopko & Daniels Sarica, 2023).

Your college may not use every item on that list. It may use a formula you cannot change. It may also have an appeal or review process that is not obvious on the testing page.

Here is the action to take today.

Placement Fact Check

Write the following six questions in your notebook. Find the answer on an official college page or ask the testing office.

1. Does the Placement Test affect admission, course placement, or both?
2. What is the exact Target Course required by my program?
3. Does the college use transcripts, prior grades, or completed courses?
4. Is corequisite support available?
5. Can I retest, and what must I do before retesting?
6. Who can review a placement that does not fit my record?

Mark every answer with one of three labels:

Confirmed means you found it on a current official source or received it from the responsible office.

Unclear means the wording could have more than one meaning.

Assumed means you have not verified it.

Do not convert Assumed into Confirmed because a forum post sounds confident.

The Cost Question

Readers often buy placement preparation because they fear losing time and money. That fear deserves honest treatment.

A lower placement may lead to additional coursework. Additional coursework may affect tuition, fees, schedule, or program timing. Those consequences vary widely. State reforms, institutional redesigns, exemptions, corequisite models, and program structures make universal cost claims unreliable.

Ask for numbers tied to your own situation:

Does the course carry degree credit?

How many credits is it?

Is it required before the Target Course?

Can it be taken at the same time?

Does it affect the next program application date?

Does financial aid cover it under my plan?

Are there lower cost review or bridge options?

Do not rely on a book to answer institutional financial aid questions. Use the book to ask better questions.

> *A vague fear grows. A priced, dated, verified consequence can be planned for.*

When a Lower Course May Help

Higher placement feels like victory because it removes a visible obstacle. Yet mathematics courses build on earlier skills. A student who enters a course without enough foundation may spend the semester repairing old gaps while trying to learn new material.

Use three tests when considering a placement result.

The requirement test: Is this course actually part of the shortest route through your program?

The readiness test: Can you perform the prerequisite skills without relying on answer recognition?

The support test: What tutoring, lab, corequisite, review, or retest options exist?

Eric, 32, needed calculus for engineering. His first placement qualified him for precalculus. He was disappointed because he had taken calculus in high school 14 years earlier. He completed a practice set without notes and found that factoring and function notation were weak. Precalculus was not evidence that he had failed. It was the course that repaired the exact tools calculus would demand.

Tasha, 41, needed introductory statistics for a social work degree. A generic test preparation site encouraged her to study advanced algebra. Her adviser showed her a quantitative pathway that did not require that sequence. Tasha's best route was not Eric's route.

Your Target Course turns "higher" into something meaningful.

What a Score Cannot Decide

A placement score cannot decide whether you belong in college.

It cannot measure your discipline after a 12-hour shift, your ability to ask for help, your knowledge of a profession, or the reason you returned.

It cannot guarantee success in a higher course.

It cannot guarantee failure in a lower one.

It cannot tell the difference between a person who never learned a concept and a person who panicked, rushed, misunderstood the interface, or prepared for the wrong test unless the broader Placement Policy makes room for that evidence.

The score is useful. It is not complete.

This distinction does not mean you should dismiss a low result. Treat the result as data. Compare it with your diagnostic, past coursework, Target Course, and available support. Then Decide.

Your One-Page Placement Brief

Before moving to Chapter 2, create a one page brief with these headings:

Institution

Program

Target Course

Placement Test or Process

Admission Consequence

Course Consequence

Multiple Measures Used

Retest Rule

Support Options

Official Contact

Date Verified

Keep this page at the front of your notebook. You will update it when you learn something new.

If a rule is missing, write "not yet confirmed." That phrase is more useful than a guess.

In a Nutshell

A Placement Test does not operate by itself. The institution's Placement Policy gives the result its consequence.

Placement may affect a starting course, a support option, or the need for further review. It may not affect admission at all. Some colleges use a single assessment heavily. Others use multiple measures, previous coursework, corequisite support, or adviser review.

The purpose of preparation is not to defeat the system. It is to give the system the clearest evidence of what you can currently do.

Confirm the consequence before you fear it. Confirm the Target Course before you chase a score. Confirm the available support before you interpret a lower placement as delay.

> ***The test produces a number. The policy produces the decision.***

Chapter 2: Find Your Exact Test

Priya ordered a placement workbook on Monday. By Wednesday, she had highlighted 40 pages of ACCUPLACER advice. On Thursday, an orientation email told her to create an ALEKS account.

She stared at the message, then at the workbook.

Both resources covered mathematics. They did not prepare her for the same experience.

ACCUPLACER mathematics questions use a multiple-choice format, and College Board says the tests have no time limit (College Board, n.d.-k). ALEKS PPL uses an adaptive, open-response assessment that can cover a broad topic range in 25 questions or fewer (McGraw Hill, n.d.-a, n.d.-b). PERT is an untimed, computer-adaptive assessment with its own operational and field-test question structure (Florida Department of Education, n.d.). TSIA2 is an untimed, computer-adaptive Texas assessment that can route students who do not meet the initial mathematics benchmark to a diagnostic test (College Board, 2020).

Priya had not wasted every minute. Fractions were still fractions. Yet she had spent scarce time learning the wrong interface, calculator expectations, and test strategy.

The chapter principle is this:

> ***Before you study mathematics, identify the exact assessment, the exact rules, and the exact route your institution uses. Content overlap does not make test formats interchangeable.***

Start With the Institution

Your college is the first source of truth.

Search the official website using combinations such as:

mathematics placement

placement testing

assessment center

testing services

new student placement

program prerequisites

retest policy

calculator policy

Add the college name to the search. Open official pages, not commercial summaries.

Record the page title and the date you checked it. Placement pages change. A screenshot from a forum may describe a policy that ended two years ago. A study guide may describe a test edition your college no longer uses.

If the page is unclear, contact the testing office or adviser. Ask one question at a time. "What do I need to do?" often produces a broad answer. "Which mathematics assessment is required for an adult applicant entering the practical nursing prerequisites in fall 2026?" produces a usable answer.

ACCUPLACER

ACCUPLACER is a College Board assessment suite used by many institutions. The mathematics tests include Arithmetic; Quantitative Reasoning, Algebra, and Statistics; and Advanced Algebra and Functions (College Board, n.d.-k).

College Board states that ACCUPLACER tests have no time limit and that the mathematics tests use multiple-choice questions (College Board, n.d.-k). Your institution can still impose appointment windows, local procedures, or additional tests.

Calculator rules are specific. College Board says you may not bring a handheld calculator for standard ACCUPLACER mathematics testing. Some questions provide an onscreen calculator. Approved accommodations may create an exception (College Board, n.d.-a).

This creates a useful practice rule. If ACCUPLACER is your test, do not complete every practice problem with your phone calculator. Learn to recognize when arithmetic should be done by hand, when estimation can check an answer, and when an onscreen tool may appear.

Official free practice resources exist. Use them after you have rebuilt the relevant skill, not only before. A sample question can diagnose a gap. A later attempt can test transfer (College Board, n.d.-c).

ALEKS PPL

ALEKS PPL stands for Placement, Preparation, and Learning. McGraw Hill describes it as an adaptive, open-response assessment that can assess knowledge across 314 topics in 25 questions or fewer (McGraw Hill, n.d.-a, n.d.-b).

Open response changes the feel of preparation. You often enter an answer rather than choose from four options. That reduces the chance of recognizing a correct answer you could not have produced independently.

Adaptive does not mean the system is punishing you. It means the assessment adjusts its questioning to estimate what you currently know.

McGraw Hill's current product page describes 12 months of access to a personalized learning module and up to five assessment attempts (McGraw Hill, n.d.-a). Because "up to five" is a vendor maximum rather than a promise from every college, confirm your institution's actual limit.

If ALEKS is your test, verify:

1. Whether the first attempt is proctored
2. Whether later attempts require time in a learning module
3. How many attempts your institution allows
4. The waiting period between attempts
5. The score needed for the Target Course
6. Whether scratch paper is permitted
7. Whether an onscreen calculator appears for selected items

Luis learned that his college allowed four attempts but required three hours in the learning module before a retest. He had planned to retake the assessment the next morning. That rule changed his schedule more than any algebra topic.

TSIA2

TSIA2 is the Texas Success Initiative Assessment 2.0. It is used to determine college readiness for many nonexempt students in Texas. Official materials explain that results may be considered with academic background, goals, and interests (College Board, 2020).

The assessment has Texas specific readiness benchmarks and diagnostic processes. Do not use a generic ACCUPLACER score chart as a substitute for current TSIA2 guidance.

College Board provides a required preassessment activity, free TSIA2 practice tests, learning resources, and downloadable sample materials (College Board, n.d.-d). Verify exemption rules and score interpretation through your institution and current Texas sources.

Daniel had an SAT score that might have affected his exemption status, but he assumed it was too old. A college adviser checked the current rule and told him exactly which documentation to submit. The useful action was not another practice set. It was finding the record.

PERT

PERT is Florida's Postsecondary Education Readiness Test. Florida's Department of Education describes it as untimed, computer adaptive, and not pass or fail. The mathematics portion includes 25 operational questions used for the placement score and five field test questions (Florida Department of Education, n.d.).

Field test questions are not labeled for the student. Do not spend the test trying to identify them. Treat every item as real.

Florida placement rules and exemptions have changed over time. Confirm whether PERT is required for your institution and situation. Some students may demonstrate readiness through other measures.

Institutional and Guided Placement

Your school may use an internally created assessment, a commercial platform with local rules, guided self placement, an adviser conversation, or previous coursework.

An institutional test can be harder to research because commercial books may not name it. Ask for:

a content outline
sample questions
time limits
allowed tools
score bands
course mappings
retest rules
review resources

A guided placement process may ask about past courses, confidence, goals, or sample problems. Treat it seriously. The goal is not to present yourself as stronger or weaker than you are. The goal is to choose a starting course with enough challenge and enough support.

Test Names Are Not Enough

Two colleges may both use ACCUPLACER and produce different course decisions from the same score because they set different Placement Policies (College Board, n.d.-i).

Two colleges may both use ALEKS and allow different numbers of attempts.

One program within a college may require a higher score than another.

The test name answers only the first question. You still need the local rule.

Heather, 39, found an old PDF that listed a score for college algebra. The college's current catalog showed a different prerequisite path. She emailed the mathematics department and learned that the old score table had been replaced after a curriculum change.

She added a new habit to her notebook: every policy fact received a date.

The Test Identification Sheet

Complete the following before moving on.

Institution:
Campus or online program:
Program:
Target Course:
Assessment name:
Assessment provider:
Sections required:
Question format:
Time limit:
Calculator rule:
Scratch paper rule:
Remote or in person:
Identification required:
Appointment required:
Number of attempts:
Waiting period:
Learning requirement before retest:
Score or placement band for Target Course:
Alternative measures:
Official page checked:
Date checked:
Office contacted:

Leave a blank beside any answer that is not confirmed. A blank is honest. An assumption is disguised uncertainty.

Practice the Interface

Once you know the test, spend one session learning the interface.

For multiple choice tests, practice eliminating impossible options without turning elimination into random guessing.

For open response tests, practice entering fractions, decimals, negative signs, exponents, and ordered pairs correctly.

For adaptive tests, practice a one question focus. Do not use the apparent difficulty of the current item to estimate your score.

For untimed tests, practice steady pacing. Untimed does not mean unlimited concentration. Fatigue still matters.

For timed tests, practice choosing when to move on.

If an official tutorial exists, use it. College Board offers free practice tests, learning resources, and downloadable sample questions for ACCUPLACER (College Board, n.d.-c).

> ***Learn the buttons before the buttons are attached to your future.***

Build a Test-Specific Practice Rule

Write one rule for your assessment.

Examples:

ACCUPLACER rule: I will practice without a handheld calculator unless a problem is marked for calculator use.

ALEKS rule: I will produce answers from a blank page instead of relying only on multiple choice recognition.

PERT rule: I will treat every item as scored and will not try to guess which questions are field test items.

Institutional test rule: I will use the official content outline and sample set, not a general test guide alone.

Place the rule at the top of every practice page.

Run the Wrong Test Cost Audit

Before buying another resource, calculate what preparing for the wrong assessment would cost you.

Write the name of the resource you are considering. Then answer:

Does it match the test used by my institution?

Does it match the sections I must take?

Does it use the same question format?

Does it reflect the current calculator rule?

Does it include topics relevant to my Target Course?

A resource can still be useful when the test name differs, but you should know which part transfers.

For example, an ACCUPLACER arithmetic lesson on fraction operations can help an ALEKS student rebuild a fraction gap. The same book's multiple choice strategy may transfer poorly to open response practice. Separate content value from format value.

Priya kept the workbook she had purchased. She relabeled it "content review" and stopped using its score charts. She used official ALEKS guidance for format and local college guidance for placement. The resource was no longer wrong. Its job had become narrower.

Use this sentence:

> *I am using this resource for __________. I am not using it to determine __________.*

A completed version might read:

"I am using this resource for algebra practice. I am not using it to determine my college's score requirement."

That boundary protects you from polished but irrelevant advice.

Confirm the Version

Test providers update names, content descriptions, interfaces, and support materials. Commercial books also publish new editions. A recent cover does not guarantee that every page reflects the current local process.

Compare the resource with:

1. the provider's current official test description
2. your institution's current placement page
3. the current course catalog

When those three sources disagree, contact the institution. Save the response or record the date and office.

A current test name attached to an old score table is still old information.

> *Current preparation begins with a date, not a cover year.*

In a Nutshell

Mathematics topics overlap across placement systems. Test experiences do not.

ACCUPLACER, ALEKS PPL, TSIA2, PERT, and local assessments can differ in question format, timing, calculators, attempts, diagnostics, and score use. Your institution adds another layer through its own Placement Policy.

Identify the test. Identify the local rule. Learn the interface. Then study.

> *The right mathematics in the wrong format is only half prepared.*

Chapter 3: Choose Your Target Course

Eric wrote "calculus" at the top of his study plan because calculus sounded like success.

He was returning for a computer science degree and remembered that technical programs required serious mathematics. He downloaded advanced algebra worksheets, watched a trigonometry review, and spent two evenings trying to factor expressions he had not seen since school.

Then he opened the degree plan.

His first required course was college algebra. Calculus appeared later and only in one concentration he had not chosen. A different track required statistics instead.

Eric had selected a prestige target, not a program target.

The chapter principle is:

> ***Your Target Course should come from the written program path, not from the highest mathematics course you can name. Preparation becomes efficient only when the destination is real.***

Find the Program Requirement

Start with the current catalog, program map, degree audit, or official advising page.

Look for the first mathematics course required by your program. Pay attention to phrases such as:

choose one of the following
prerequisite
corequisite
minimum grade
placement required
statistics pathway
quantitative reasoning
college algebra
technical mathematics
calculus sequence

Do not stop at the course title. Open the course description and read the prerequisites.

A course may require a specific placement band, completion of another course, or permission from a department. A program may accept several mathematics options. One option may fit your goals better than another.

Tasha was entering a social work program. She assumed "college math" meant college algebra. The program accepted introductory statistics, which connected more directly to later research methods. Her Target Course changed from algebra to statistics.

Carlos was entering an industrial maintenance certificate. His program required technical mathematics built around measurement, formulas, and applied algebra. A generic college algebra plan would have spent too much time on topics outside his first course.

The title on the test does not choose the pathway. The program does.

Build the Course Chain

A Target Course rarely stands alone. Draw the chain that leads into it and the courses that follow it.

For example:

Foundation course College algebra Precalculus Calculus I

Or:

Corequisite support plus statistics Research methods

Or:

Technical mathematics Applied physics

Write the chain as your institution presents it. Do not use a chain from another college.

Now circle the first course you need. That is the Target Course.

Then underline the immediate prerequisite skills. Those skills shape your High-Value Skills list.

If your Target Course is statistics, you may need strong fractions, percentages, ratios, signed numbers, formulas, graphs, and basic algebra. If it is calculus, you will need a broader algebra and function foundation. If it is dosage calculation within a health program, the institution may require a specific prerequisite course and later apply mathematics in a clinical context. This book can refresh ratios and conversions, but it does not replace program instruction or clinical competency checks.

Eligibility Is Not Readiness

Placement systems often answer, "May this student enroll?"

You must also ask, "Can I begin this course with a reasonable chance of success?"

Those questions overlap. They are not identical.

Renee barely met the placement score for college algebra. She could solve one step equations but struggled with fractions inside equations. The score made her eligible. Her No-Shame Diagnostic showed a Rebuild Skill that the course would use immediately.

She had three reasonable choices:

1. Delay and complete a review course
2. Enroll with tutoring and a structured weekly review
3. Use the time before the semester to repair that gap

The right choice depended on time, course design, support, and the rest of her schedule.

A higher score can open a door. It cannot carry you through the room.

Placement is permission. Readiness is preparation.

Ask What Success Requires

Course descriptions often list content but not the pace.

Ask an adviser, instructor, tutoring center, or department:

What skills do successful students use in the first three weeks?

Is there a readiness checklist?

Is a review module available before class?

How many hours of work are typical outside class?

Is tutoring built into the course?

Is a corequisite support section available?

Does the course use a graphing calculator or software?

Does my program require a particular grade?

These questions reveal more than a cut score alone.

Omar qualified for statistics but learned that the section he wanted was compressed into eight weeks. He worked full time and had not taken mathematics in 17 years. He chose a 16 week section instead. His placement did not change. His chance of completing the course did.

Separate Required From Optional

Create three columns.

Required Now

List the topics needed for the Placement Test and Target Course.

Useful Later

List topics that may matter in later courses but are not urgent.

Not Yet Confirmed

List anything you think may matter but have not verified.

This simple division prevents advanced topics from consuming time needed for basic ones.

Mei enjoyed reviewing geometry because it felt familiar. She avoided equations because they felt uncomfortable. Her test outline emphasized algebra. Geometry was useful later. Equations were required now.

A study plan can become a hiding place. We often review what feels productive, not what has the highest value.

The High-Value Skill is not always the pleasant skill.

Three Route Types

This book uses three preparation routes.

Foundation Route

Choose this route when the diagnostic shows Rebuild Skills in whole number operations, signed numbers, fractions, decimals, percentages, or basic equation meaning.

The Foundation Route prioritizes Chapters 6 through 10. It uses shorter problem sets and more repeated review.

College-Ready Route

Choose this route when arithmetic is mostly stable but algebra, word problems, graphs, or mixed problem selection need work.

The College-Ready Route prioritizes Chapters 7 through 12, then moves quickly into Mixed Practice.

Higher-Math Route

Choose this route only when your Target Course or official test outline requires advanced algebra, functions, geometry, or precalculus topics.

The Higher-Math Route includes Chapter 13 after the core chapters. It is not an award for being “good at math.” It is a route requirement.

You may use a blended route. A reader can be Higher-Math in one topic and Foundation in another. The route tells you where to spend time, not who you are.

The Target Course Map

Complete this worksheet.

Program name:
Catalog year:
Required mathematics course or options:
My selected Target Course:
Course number and title:
Credits:
Prerequisite course or placement:
Minimum grade required:
Courses that follow:
Program deadline affected by this course:
Skills listed in the course description:
Skills named by an adviser or readiness guide:
Support available:
Date verified:

Now complete this sentence:

> *My goal is to demonstrate readiness for __________, not to chase __________ simply because it sounds higher.*

For Eric, the sentence became:

"My goal is to demonstrate readiness for college algebra, not to chase calculus simply because it sounds higher."

That sentence saved him days.

Course Choice And Career Goals

Some programs permit several mathematics courses. Choose with guidance, not only comfort.

A statistics course may suit a program that emphasizes data interpretation. College algebra may be required for later science or calculus. Quantitative reasoning may satisfy a general education requirement but not a technical prerequisite.

Ask what the course unlocks.

Aisha's nursing prerequisite list allowed either college algebra or statistics, but a later transfer program preferred statistics. She checked the transfer agreement before choosing.

Luis needed mathematics for a skilled trades program and planned to move into supervision later. His adviser recommended a technical course first and explained how a later general education course could support the degree route.

The best Target Course solves the next real problem without closing a likely future path.

The Fantasy Score Trap

A fantasy score has no verified purpose.

It may come from:

a forum user's result

a competitor book's promise

an old score chart

a desire to prove intelligence

fear of being judged

the assumption that higher is always cheaper

Replace the fantasy score with a course requirement and a readiness standard.

If the institution gives a score band, record it. If it uses guided placement, record the process. If it uses multiple measures, record each measure you can influence or document.

Then create two goals:

Placement Goal: The course outcome you want.

Readiness Goal: The skills you want to perform accurately before the first class.

For example:

Placement Goal: College algebra.

Readiness Goal: Solve linear equations, work accurately with fractions and signed numbers, interpret graphs, and translate basic word problems.

The second goal keeps helping after the test is over.

Action Challenge

Find one official source and one human source.

The official source should be the current catalog, program map, or placement page.

The human source should be an adviser, testing staff member, department representative, or tutor familiar with the course.

Ask both sources the same question:

> ***What mathematics course should I enter first for this program, and what skills will I need in the opening weeks?***

Record any difference between the answers.

If they conflict, do not average them. Ask which source controls the decision.

Compare Two Honest Routes

Suppose a program accepts either college algebra or quantitative reasoning.

Route A enters college algebra, which leads into precalculus and later calculus.

Route B enters quantitative reasoning, which satisfies the degree requirement but does not open the calculus chain.

Neither route is automatically better.

If you plan to transfer into engineering, Route A may protect a later prerequisite. If your degree requires one general education mathematics course and your field emphasizes reasoning with data, Route B may fit more directly.

Now add readiness.

A student with strong algebra skills may enter Route A confidently. A student with fragile equation skills, a full work schedule, and no later calculus requirement may complete Route B more efficiently. The decision must account for both future doors and present preparation.

Use a two column comparison:

Route A

What it satisfies What it unlocks What it requires now Time and support demands

Route B

What it satisfies What it unlocks What it requires now Time and support demands

Take that comparison to an adviser. Do not ask, "Which course is easier?" Ask, "Which course fits this program, this transfer plan, and my demonstrated preparation?"

Define Success Before the Score

Write three descriptions.

Minimum useful placement

The lowest outcome that still allows progress without an unnecessary delay.

Preferred placement

The outcome that matches the Target Course and current readiness.

Stretch placement

A higher outcome that would be useful only if prerequisite skills are reliable.

This protects you from treating every result below the stretch as failure.

Eric's minimum useful placement was precalculus because it counted toward his technical pathway. His preferred placement was college algebra only until he confirmed the actual sequence. After advising, his preferred placement became precalculus. The facts changed the label.

A score goal without a course map is a number looking for meaning.

Confirm the Course Number

Course titles can look similar while serving different pathways. "College Mathematics," "Quantitative Reasoning," and "College Algebra" may carry different numbers, prerequisites, and transfer uses.

Record the exact course number, not only the title.

If an adviser says "take math 110," compare that number with the current catalog and your program map. If the program accepts several options, ask which one transfers or unlocks later science courses.

A one digit difference can represent a different route.

Your Target Course is specific enough to register for. If you cannot name the course number or official placement outcome, the target still needs clarification.

In a Nutshell

A useful preparation plan begins with a real destination.

Find the current program requirement. Trace the prerequisite chain. Select the first course you actually need. Then separate eligibility from readiness.

Your Target Course determines which topics are High-Value Skills and which can wait.

The highest named course is not the highest value outcome.

> ***Study for the course that moves your program forward, then prepare well enough to stay there.***

Chapter 4: Take the No-Shame Diagnostic

James printed the diagnostic at 9:10 on a Saturday morning.

By 9:16 he had checked the answer to question three.

He told himself he only wanted to confirm that he understood the directions. Five minutes later he had checked question seven. Then question 11. By the time he finished, the score looked reassuring and told him almost nothing.

James had turned a diagnostic into a performance.

On Sunday he tried again with a different short set. This time he wrote a question mark beside anything he could not begin within 30 seconds. He completed the rest without notes. He did not guess. He showed every step.

His score was lower. His plan was better.

The chapter principle is:

> *The No-Shame Diagnostic is not designed to make you feel ready. It is designed to show you what to do next. An honest blank is more useful than a lucky answer.*

The No-Shame Diagnostic is a study-planning tool created for this workbook. It is not a validated placement assessment, it cannot predict an institution's course decision, and its Ready, Recall, and Rebuild labels should be used only to organize review.

Set Up the Conditions

Take the diagnostic in Appendix A before studying the mathematics chapters.

Use:

pencil or pen

plain scratch paper

no notes

no search engine

no answer key

no handheld calculator unless your verified test rules make it appropriate

The diagnostic is untimed, but note the start and finish time. Do not hurry. Do not turn it into a three hour endurance test either. If you are stuck for more than two minutes, mark the question and continue.

Write one of four symbols beside each answer:

C means confident.

U means unsure.

P means you needed a prompt from memory, such as drawing a model or recalling a similar example.

B means blank.

These symbols matter because a correct answer with low confidence may still need review. A wrong answer with a sound method and one arithmetic slip may require a different response from a complete blank.

Do Not Guess for Pride

Multiple choice tests may reward strategic elimination. A diagnostic has a different job.

If you guess correctly, you can hide a Rebuild Skill.

Suppose the question asks:

$$\frac{3}{4}+\frac{1}{6}$$

You see four choices and select $\frac{11}{12}$ because it "looks right." The answer is correct. Yet if you cannot explain the common denominator, the skill is not Ready.

During the diagnostic, produce the answer from your own steps.

If you cannot, mark the question. That is not failure. That is routing information.

> *A blank today can prevent a surprise on test day.*

Score by Skill Group

The diagnostic is divided into six domains:

1. Number sense and signed numbers
2. Fractions, decimals, and percentages
3. Ratios, rates, and units
4. Algebraic expressions and equations
5. Graphs, data, and functions
6. Word problems and higher mathematics

Do not look only at the total.

A total score can hide an uneven profile.

Mei answered 22 of 30 questions correctly. That sounded strong. Her misses were concentrated in signed numbers and equations. Those topics appeared throughout her Target Course.

Carlos answered 18 correctly. His errors were spread across advanced items outside his technical mathematics target. His core arithmetic and measurement skills were strong.

Mei needed a focused rebuild despite the higher total. Carlos needed a narrower review despite the lower total.

The Placement Gap is not simply the distance between your score and perfection. It is the distance between your current performance and the skills needed for your Target Course.

Classify Every Skill

After checking the answers, classify each skill.

Ready Skill

Mark a skill Ready when you can:

solve the problem accurately
explain the method
repeat the method on a similar problem
check whether the answer is reasonable

One correct answer is evidence. It is not always enough evidence.

Recall Skill

Mark a skill Recall when:

a short example brings the method back
you understand the explanation quickly
errors decrease after a small number of problems
the concept feels familiar but slow

Recall Skills often improve quickly, but they fade if you review once and never return.

Rebuild Skill

Mark a skill Rebuild when:

you do not know how to begin
the explanation uses unfamiliar language
you repeat the same type of error
an earlier missing skill blocks the topic
you can copy a method but cannot use it on a changed problem

A Rebuild Skill deserves time and instruction. It does not deserve shame.

Find the First Broken Link

Mathematics errors often sit downstream from another gap.

A person may struggle with equations containing fractions because fraction operations are weak. They may struggle with percentages because decimals are unstable. They may struggle with slope because signed numbers and division are uncertain.

When you miss a problem, ask:

1. What was the final skill being tested?
2. What smaller skills did it require?
3. Which step was the first place I became unsure?

That first uncertain step is often the First Broken Link.

Daniel missed:

$$3(x - 2) = 15$$

He knew he should "do something with the three" but divided only the 15 and forgot that the expression $x - 2$ was being multiplied as a group. His First Broken Link was not subtraction. It was understanding multiplication around parentheses.

Renee missed:

$$\frac{x}{4} + 3 = 8$$

She subtracted three correctly, then multiplied five by four incorrectly. Her equation method was Ready. Her basic multiplication error was an Error Pattern, not a missing algebra concept.

Different causes need different repairs.

Start the Error Pattern Log

Create four columns.

Problem

Write the question number or a short description.

What I Did

Record the actual step, not "careless mistake."

Error Pattern

Choose one:

sign error
operation choice
fraction rule
decimal placement
copied number
vocabulary
formula selection
skipped step
rushed reading
unknown concept

Prevention Rule

Write a specific action.

Examples:

"Circle every negative sign before calculating."
"Write the units beside every number."
"After solving, substitute the answer."
"Do not add denominators."
"Read the final sentence twice."

Your log should not say "be more careful." Careful is not a procedure.

Name the error so you can interrupt it.

Route Yourself

Use the full routing chart in Appendix B. The basic rule is:

Choose the **Foundation Route** when you have Rebuild Skills in basic operations, signed numbers, fractions, decimals, percentages, or one step equations.

Choose the **College-Ready Route** when those foundations are mostly Ready but you need work on algebra, graphs, functions, data, word problems, or mixed selection.

Choose the **Higher-Math Route** only when the core skills are stable and your Target Course requires the advanced topics in Chapter 13.

A blended route is allowed. For example:

Foundation in fractions College-Ready in graphs Higher-Math in exponents

The 30-day calendars tell you how to combine routes.

Retest One Skill Immediately

After scoring, choose one Recall Skill.

Read the relevant explanation in this book. Close the book. Solve two fresh problems.

If both are correct and you can explain them, the Recall label is probably accurate.

If you can only repeat the exact example, change the label to Rebuild.

This small test protects you from false familiarity.

Heather remembered how to convert 0.35 to 35 percent as soon as she saw one example. She then converted 0.072 to 7.2 percent without help. Recall was accurate.

Omar copied the steps for adding $\frac{1}{3}$ and $\frac{1}{4}$. When the next problem used $\frac{2}{5}$ and $\frac{3}{10}$, he added the denominators. The skill needed rebuilding.

When the Diagnostic Feels Bad

Mathematics anxiety has a documented negative relationship with mathematics performance across many studies (Barroso et al., 2021). That does not mean every anxious feeling will lower every score. It means the feeling is common enough, and connected strongly enough to performance, that your study process should account for it.

Use a two minute reset:

1. Put both feet on the floor.
2. Exhale slowly for longer than you inhale.
3. Read the problem once without solving.
4. Underline the quantity being requested.
5. Write one fact you know.
6. Begin with the smallest available step.

This is not a cure for anxiety. It is a way to restart action.

If distress is severe, persistent, or connected to a broader health concern, use the support available through your institution or a qualified professional. A study guide is not a substitute for care.

Diagnostic Reflection

Answer these prompts.

The domain with the most Ready Skills is:

The domain with the most Recall Skills is:
The domain with the most Rebuild Skills is:
My most common Error Pattern is:

> *The skill most important to my Target Course is:*

The first chapter I need is:
The first external support I may need is:
Then write:

> *My diagnostic result is a set of instructions, not a description of my intelligence.*

Separate Knowledge From Process

After scoring, create two totals.

Content errors happen when the concept or method is missing.

Process errors happen when you know the method but lose accuracy through reading, signs, copying, entry, or checking.

The distinction matters.

If eight misses are content errors, you need instruction.

If eight misses are process errors, rereading the same lesson may not help. You need a stronger scratch work system, a reading routine, or an answer check.

James originally classified every miss as "I do not know math." His second review found three unknown concepts and five process errors. The plan changed from eight broad review sessions to three content lessons and five prevention rules.

For each incorrect answer, complete this sentence:

> *I would have answered correctly if I had known or done* __________.

Examples:

"I would have answered correctly if I had known how to find a common denominator."

"I would have answered correctly if I had copied the negative sign."

"I would have answered correctly if I had read that the question asked for the new price."

The first needs instruction. The second and third need process repair.

Use a Two-Day Confirmation

Do not finalize a Ready label immediately after reading an explanation.

Day one:

Study the method and solve two problems.

Day two:

Solve one fresh problem without notes.

If the skill survives the delay, keep Ready or Recall.

If it disappears, change it to Rebuild or schedule another retrieval.

This two day confirmation prevents a common illusion. Right after an example, the method is still visible in short term memory. The delayed problem asks whether you can retrieve and adapt it.

Protect the Diagnostic Record

Keep the original diagnostic untouched.

Do not erase wrong work or replace blanks with corrected answers. Add corrections on a separate page.

The original is valuable evidence. At the end of 30 days, solve selected items again and compare the process, not only the score.

Look for:

shorter hesitation

clearer scratch work

fewer repeated Error Patterns

stronger explanation

better estimates

correct answer form

Progress is easier to trust when you can see the page where it began.

Mark Uncertain Correct Answers

Do not ignore a correct answer marked U.

Choose two uncertain correct answers and solve fresh versions the next day. If both remain correct with clear work, change the skill to Ready. If the method disappears, classify it Recall or Rebuild.

Correctness tells you what happened once. Confidence and delayed performance help tell you how reliable the skill is.

In a Nutshell

Take the diagnostic honestly.

Do not use the answer key, guess for pride, or judge yourself by the total alone. Score by domain. Classify each skill as Ready, Recall, or Rebuild. Identify the First Broken Link and begin an Error Pattern Log.

Your starting score is not the product you came here to create. Your study route is.

> *The most useful diagnostic answer is the one that tells the truth soon enough to change the plan.*

Chapter 5: Build a Plan That Fits

Omar's first study schedule looked excellent on paper.

He planned to work for 90 minutes every evening after his shift. He bought three notebooks, arranged highlighters by color, and wrote "NO EXCUSES" across the top of the calendar.

On day one, his shift ran late. On day two, his daughter needed help with a school project. On day three, he sat down at 10:20 p.m., read the same paragraph four times, and closed the book.

By day five, the empty boxes on the calendar felt like evidence that he lacked discipline.

Omar's problem was not discipline. His plan required the same energy every day from a life that did not produce the same energy every day.

He replaced the 90 minute promise with three versions of success:

15 minutes on a hard day. 30 minutes on a normal day. 45 minutes when time and energy allowed.

He also added two catch up windows each week. The new plan looked less impressive. He followed it.

The chapter principle is:

> ***The best 30-day plan is not the most ambitious plan. It is the plan that still functions on the week your life behaves like your life.***

Choose Your Daily Route

Use the diagnostic and your schedule to select one of three time routes.

The 15-Minute Route

Use this when work, caregiving, health, or time makes longer sessions unreliable.

A 15 minute session follows this structure:

Two minutes: recall yesterday's rule without notes.

Eight minutes: study one small skill and solve two or three problems.

Three minutes: check answers and update the Error Pattern Log.

Two minutes: write the first problem for tomorrow.

The final step reduces starting friction. When you return, the page already tells you what to do.

The 30-Minute Route

Use this as the standard route.

Five minutes: retrieval from an earlier day.

15 minutes: instruction and focused practice.

Five minutes: Mixed Practice.

Five minutes: answer check and Error Pattern Log.

This route gives enough time to learn one focused idea without turning the session into a broad survey.

The 45-Minute Route

Use this when your Placement Gap is larger, your deadline is firm, or your Target Course requires Higher-Math topics.

Five minutes: retrieval.

20 minutes: instruction and worked examples.

10 minutes: independent practice.

Five minutes: Mixed Practice.

Five minutes: error review and plan for the next session.

Stop after 45 focused minutes unless you are still working well. Longer is not automatically better. Fatigue can turn practice into repeated error.

Build Three Kinds of Days

A 30-day plan should not treat every day the same.

Learning Days introduce or rebuild a skill.

Review Days return to earlier skills and mix them.

Assessment Days use diagnostic or test style questions without immediate help.

A simple pattern is:

Three learning days One review day Two learning days One catch up or rest day

Repeat the rhythm. In the final week, shift more time toward Mixed Practice and the Calm Test Routine.

In a college calculus study, spaced retrieval practice improved end-of-semester retention even though performance was lower on the first two spaced practice questions than on comparable massed questions (Lyle et al., 2022). That uncomfortable return is not proof that the method failed. It is the moment memory has to work.

Use the Minimum Viable Session

The Minimum Viable Session is what you do when the planned session is not possible.

It has three steps:

1. Solve one problem from an earlier topic.
2. Review one Error Pattern.
3. Write the next action.

This can take five minutes.

It does not replace regular study. It protects continuity.

Heather missed a full session because of a family appointment. Before bed, she solved one percentage conversion, reviewed her decimal placement rule, and opened the book to the next lesson. The next day she resumed without the mental weight of "starting again."

A small session keeps the door open.

Plan Around Energy

Time is not the only resource. A 30 minute block at 6:30 a.m. may be more useful than an hour at 11:00 p.m.

For three days, record when you feel most able to concentrate.

Look for:

a quiet period before work

a lunch break

time while a child is at an activity

the first 30 minutes after arriving home

a weekend morning

a library hour

a commute suitable for vocabulary review, but not written calculation

Protect demanding topics for higher energy periods. Use lower energy periods for answer review, vocabulary, formula cards, or organizing the Error Pattern Log.

Do not choose a study time because it sounds disciplined. Choose it because it repeatedly exists.

The Missed Day Rule

When you miss a day, do not restart the 30-day plan.

Use this rule:

Miss one day: Continue with the next scheduled session and move the missed work to a catch up window.

Miss two days: Complete a 15 minute review before starting new material.

Miss three or more days: Retest three representative skills, then adjust the calendar.

Do not compress four missed sessions into one exhausted evening.

Priya lost three days to illness. She tried to double every remaining session and quickly began making errors she had already fixed. She returned to the route chart, dropped two low priority topics, and kept the High-Value Skills.

A deadline may be fixed. The plan must still be edited.

Decide What to Drop

A focused study plan needs a stop list.

Drop or delay a topic when:

it is outside the official test outline

it is not a prerequisite for the Target Course

it is a Higher-Math topic your route does not require

it consumes time needed for a First Broken Link

it has already become a Ready Skill

Do not keep solving 20 percentage problems because percentage problems feel good. Move to the skill that still changes the placement decision.

Likewise, do not spend three days on one advanced problem if the test gives that topic little weight and your arithmetic foundation remains weak.

> *Study value is not the same as study difficulty.*

Know When to Get Help

Independent study works best when you can identify and correct errors.

Seek outside help when:

the same Rebuild Skill remains unclear after two focused sessions

answer explanations use vocabulary you do not understand

you cannot identify the First Broken Link

you repeat an Error Pattern after applying a prevention rule

your deadline is close and the skill is central to the Target Course

anxiety prevents you from beginning or completing practice

Possible sources include a college tutoring center, adult education program, mathematics lab, instructor office hour, reputable online lesson, or private tutor.

Bring a specific question.

Weak question: "Can you help me with algebra?"

Strong question: "I can solve $3x + 5 = 20$, but I do not understand what to do when the variable appears on both sides."

The stronger question saves time and reveals the gap.

The 30-Day Structure

The appendices include detailed calendars. The overall structure is:

Days 1 through 3: Clarify and Diagnose

Confirm the test, Placement Policy, Target Course, and diagnostic route.

Days 4 through 9: Number Foundations

Work on signed numbers, fractions, decimals, percentages, ratios, and units according to your route.

Days 10 through 17: Algebra Core

Translate expressions, simplify, solve equations, work with inequalities, and use formulas.

Days 18 through 22: Graphs and Word Problems

Read tables, graphs, functions, and multistep situations.

Days 23 through 25: Route Specific Work

Use Higher-Math topics if required, or repair remaining Foundation gaps.

> *Days 26 through 27: Practice Assessment One and Repair*

Complete a mixed assessment. Classify every error. Review only the gaps the assessment reveals.

> *Days 28 through 29: Practice Assessment Two and Test Routine*

Use realistic conditions and rehearse logistics.

Day 30: Light Review and Readiness Check

Do not attempt to learn an entire new domain. Review your prevention rules, formula notes, test details, and Calm Test Routine.

Create Backup Windows

Write your main study time and two backups.

Example:

Main: Monday through Friday, 7:00 to 7:30 a.m.

Backup one: Lunch break, Tuesday and Thursday.

Backup two: Saturday, 9:00 to 10:00 a.m.

A backup window is not free time waiting to be discovered. Put it on the calendar.

Jordan used Saturday morning as a catch up block. When no catch up was needed, he used it for Mixed Practice. The time never became a punishment.

Use Visible Evidence

Confidence should come from evidence you can see.

Track:

sessions completed

Ready Skills gained

Recall Skills recovered

Rebuild Skills still active

Error Patterns reduced

practice assessment results

unanswered policy questions

Do not track only hours. Two focused 20 minute sessions can be more valuable than an hour of rereading.

At the end of each week, write:

This week I can now do:

The error I am making less often is:

The next High-Value Skill is:

That is evidence based confidence.

Your Calendar Action

Turn to Appendix C and choose the 15, 30, or 45 minute calendar.

Write your test date.

Count backward 30 days. If you have fewer than 30 days, keep the order but combine only compatible sessions. Do not remove assessment days.

If you have more than 30 days, add review spacing. Do not fill every extra day with new topics.

Mark:

work peaks

caregiving commitments

appointments

travel

known low energy days

backup windows

assessment days

Then sign this sentence:

> ***I will follow the smallest version of the plan on difficult days rather than abandoning the plan because the largest version is unavailable.***

Design the First Five Minutes

Many study plans fail before the mathematics begins. The learner spends the available time deciding what to do, finding a pencil, opening three websites, and choosing a topic.

End every session by preparing the first five minutes of the next one.

Write the date.

Write the skill.

Copy the first question.

Place the needed page marker.

Close unrelated tabs.

Then the next session begins with action rather than negotiation.

Omar's first prepared question was:

$$\frac{2}{3} + \frac{1}{4}$$

The next morning, he did not need to decide whether he felt ready for fractions. The problem was already waiting.

Use a Stop Rule

A session also needs an ending.

Stop when:

the planned time is complete

two consecutive errors show fatigue rather than a new gap
you can state the next action
the current problem has been checked and logged

Do not end in the middle of a confused page when five more minutes can identify the First Broken Link. Do not continue for 40 extra minutes because one set went well.

A reliable stop rule protects tomorrow's session.

Pair Difficult and Familiar Work

On low confidence days, begin with one Ready Skill, then move to a Rebuild Skill, then finish with one Mixed Practice problem.

The familiar problem creates movement. The Rebuild work remains the center. The final mixed item tests selection.

This is not avoiding difficulty. It is arranging difficulty so the session has an entrance and an exit.

Write tomorrow's sequence now:

Ready Skill:

Rebuild Skill:

Mixed Practice item:

The calendar tells you when to study. This sequence tells you how to enter the work.

In a Nutshell

Your study schedule must survive ordinary disruption.

Choose a 15, 30, or 45 minute route. Use Learning Days, Review Days, and Assessment Days. Protect continuity with the Minimum Viable Session. Use catch up windows instead of restarting.

Study High-Value Skills. Drop low priority work. Ask for help with a specific problem when independent review stops producing change.

> *A plan is reliable when it tells you what to do on the bad day, not only the good one.*

Chapter 6: Reset Your Number Sense

A correct method can still end in a wrong answer when the final calculation ignores sign, size, or operation. Number sense is the check that catches that mismatch before you move on.

> ***Number sense is not fast mental arithmetic. It is the ability to understand size, sign, operation, and reasonableness well enough to catch an answer that cannot be right.***

Begin With Place Value

A digit's value depends on its position.

In $4,582.73$:

4 represents four thousands

5 represents five hundreds

8 represents eight tens

2 represents two ones

7 represents seven tenths

3 represents three hundredths

Placement tests often use place value indirectly. You may need to compare decimals, round a number, estimate a result, or position a decimal point after multiplication or division.

Compare 0.8 and 0.75.

Write 0.8 as 0.80. Now the comparison is 0.80 versus 0.75. Eighty hundredths is greater than 75 hundredths.

A common error is to assume the number with more digits is larger. That rule works poorly with decimals. 0.125 has more digits than 0.8, but 0.125 is smaller.

Quick Check

1. Which is larger, 0.4 or 0.39?
2. Round 18.746 to the nearest tenth.
3. What is the value of the 6 in 3.064?

Answers:

1. $0.4 = 0.40$, so 0.4 is larger.
2. 18.7, because the hundredths digit is 4.
3. Six hundredths, or 0.06.

Estimate Before Calculating

Estimation is not a substitute for an exact answer when an exact answer is required. It is a safety check.

Suppose you calculate:

$$49.8 \times 6.1$$

Round to $50 \times 6 = 300$.

If your exact result is 30.378, the decimal point is wrong. If it is 303.78, the result is reasonable.

Try another:

$$812 \div 19$$

Round to $800 \div 20 = 40$.

An exact result near 43 is reasonable. A result of 4.3 or 430 should trigger a check.

Before every multistep problem, ask:

About how large should the answer be?

That question can catch mistakes before the answer key does.

Renee used estimation on a tuition discount problem. The original cost was about $2,000 and the discount was 15 percent. She knew the discount should be a few hundred dollars, not $30 and not $3,000. Her estimate helped her reject an answer produced by a misplaced decimal.

Use the Four Operations Deliberately

Addition combines quantities.

Subtraction finds a difference or removes a quantity.

Multiplication represents repeated groups, scaling, or area.

Division separates into equal groups, finds how many groups fit, or creates a rate.

Word problems become easier when you name the relationship before touching the numbers.

Example:

A course requires 48 hours of lab work completed over 12 weeks. How many hours per week?

The total is divided into 12 equal weekly groups:

$$48 \div 12 = 4$$

Answer: four hours per week.

Example:

A worker earns \$18.50 per hour for 7 hours. What is the gross pay for those hours?

The hourly amount is repeated seven times:

$$18.50 \times 7 = 129.50$$

Answer: \$129.50.

Units clarify the operation. Dollars per hour multiplied by hours leaves dollars.

Follow the Order of Operations

Use this order:

1. Parentheses and grouping
2. Exponents
3. Multiplication and division from left to right
4. Addition and subtraction from left to right

The phrase "from left to right" matters.

Evaluate:

$$18 \div 3 \times 2$$

Division and multiplication have equal priority. Work left to right:

$$18 \div 3 = 6$$

$$6 \times 2 = 12$$

Do not multiply 3×2 first unless parentheses tell you to.

Evaluate:

$$4 + 3(6 - 2)^2$$

Parentheses:

$$6 - 2 = 4$$

Exponent:

$$4^2 = 16$$

Multiplication:

$$3 \times 16 = 48$$

Addition:

$$4 + 48 = 52$$

Heather's Error Pattern was skipping the exponent when a problem also contained parentheses. Her prevention rule was to copy the expression onto a new line after each stage. The extra lines made her work easier to check.

Understand Positive and Negative Numbers

A positive number lies to the right of zero. A negative number lies to the left.

For addition:

Same signs: add the absolute values and keep the sign.

Different signs: subtract the smaller absolute value from the larger and keep the sign of the number with the larger absolute value.

Examples:

$$-7 + (-5) = -12$$
$$-9 + 4 = -5$$
$$8 + (-3) = 5$$

For subtraction, add the opposite:

$$6 - (-4) = 6 + 4 = 10$$
$$-3 - 7 = -3 + (-7) = -10$$

For multiplication and division:

Same signs produce a positive result.

Different signs produce a negative result.

Examples:

$$(-4)(-6) = 24$$
$$(-4)(6) = -24$$
$$36 \div (-9) = -4$$

Do not memorize sign rules without checking meaning. If you owe \$7 and then owe \$5 more, your position is -12. If a temperature rises four degrees from -9, it becomes -5.

Use Absolute Value Correctly

Absolute value is distance from zero.

$$|-8| = 8$$
$$|5| = 5$$

Distance is nonnegative.

This does not mean “remove every negative sign.” Consider:

$$-|-8| = -8$$

The absolute value of -8 is 8. The negative sign outside the bars remains, producing -8.

Placement questions may compare absolute values or use them in expressions. Slow down and identify what is inside the bars.

Factors Multiples and Divisibility

A factor divides a number evenly.

The positive factors of 12 are:

$$1, 2, 3, 4, 6, 12$$

A multiple is produced by multiplying.

Multiples of 6 include:

$$6, 12, 18, 24, 30$$

Factors help with simplifying fractions and factoring algebraic expressions. Multiples help with common denominators.

Useful divisibility checks:

divisible by 2 if the last digit is even

divisible by 3 if the digits add to a multiple of 3

divisible by 5 if the last digit is 0 or 5

divisible by 10 if the last digit is 0

Example: Is 372 divisible by 3?

$$3 + 7 + 2 = 12$$

Since 12 is divisible by 3, 372 is divisible by 3.

Write Work That Can Be Checked

Scratch work is not evidence that you are weak. It is an external memory system.

Use one line for one transformation.

Poor layout:

$$3 + 4(7 - 2) = 3 + 20 = 23$$

This answer happens to be correct, but the compressed line is harder to inspect.

Clear layout:

$$3 + 4(7 - 2)$$
$$= 3 + 4(5)$$
$$= 3 + 20$$
$$= 23$$

When a problem contains negative signs, circle them. When it contains decimals, align decimal points in addition and subtraction. When it contains units, carry the units through.

A testing screen may not score your scratch work. The work still protects the answer.

The Number Sense Reset

Complete these without a calculator.

1. $4,806 + 975$
2. $7,000 - 2,468$
3. 36×14
4. $864 \div 24$
5. $5 + 3(8 - 6)^2$
6. $-7 + 12$
7. $9 - 15$
8. $(-6)(-8)$
9. $45 \div (-5)$
10. Which is larger, 0.607 or 0.67?
11. Round 92.485 to the nearest hundredth.
12. Estimate 39.7×5.2.

Answers:

1. $5,781$
2. $4,532$
3. 504
4. 36
5. 17
6. 5
7. -6
8. 48
9. -9
10. 0.67, because $0.670 > 0.607$
11. 92.49
12. About $40 \times 5 = 200$

For every error, record the first wrong step.

The Solve Check Explain Routine

Use this routine on the final four problems.

Solve: Show the calculation.

Check: Estimate, reverse the operation, or substitute.

Explain: Write one sentence naming the rule.

Example:

$$864 \div 24 = 36$$

Check:

$$36 \times 24 = 864$$

Explain: Division and multiplication are inverse operations.

This routine feels slower during practice. It builds the habit that catches preventable mistakes during testing.

Use Inverse Checks

Several operations can be checked by reversing them.

Addition:

$$468 + 257 = 725$$

Check:

$$725 - 257 = 468$$

Subtraction:

$$902 - 487 = 415$$

Check:

$$415 + 487 = 902$$

Multiplication:

$$36 \times 14 = 504$$

Check:

$$504 \div 14 = 36$$

Division:

$$864 \div 24 = 36$$

Check:

$$36 \times 24 = 864$$

You do not need to reverse every simple calculation on test day. Use inverse checks when the result affects several later steps or when your Error Pattern Log shows repeated arithmetic errors.

Compare Before You Calculate

Comparison can simplify work.

Which is larger?

$$\frac{7}{8}$$

or:

$$0.82$$

Convert or estimate.

$$\frac{7}{8} = 0.875$$

So $\frac{7}{8}$ is larger.

Which is closer to 50?

47.8 or 52.6?

Distances:

$$50 - 47.8 = 2.2$$
$$52.6 - 50 = 2.6$$

So 47.8 is closer.

Number sense includes distance and relative size, not only operations.

Build Five Benchmarks

Memorize useful benchmarks:

$$\frac{1}{2} = 0.5$$
$$\frac{1}{4} = 0.25$$
$$\frac{3}{4} = 0.75$$
$$10\% = 0.1$$
$$100\% = 1$$

Use them to estimate.

Thirty percent of 90 should be less than half of 90 and greater than 10 percent of 90. An answer of 27 fits.

A result of 270 does not.

Benchmarks give your exact calculation a guardrail.

Use Rounding Without Losing the Question

Rounding can estimate, but do not replace an exact answer unless the question requests it.

Suppose a program cost is $1,987 and you want an estimate of 12 percent.

Round to $2,000:

$$0.12(2,000) = 240$$

The exact amount is:

$$0.12(1,987) = 238.44$$

The estimate predicts the neighborhood. The exact calculation supplies the answer.

When rounding a final answer, follow the requested place.

Nearest whole number: inspect the tenths digit.

Nearest tenth: inspect the hundredths digit.

Nearest hundredth: inspect the thousandths digit.

For money, two decimal places are usually used unless the problem says otherwise.

Check Division With Size

If you divide by a number greater than one, a positive result should usually be smaller than the original positive number.

$$84 \div 7 = 12$$

If you divide by a positive number between zero and one, the result becomes larger.

$$12 \div 0.5 = 24$$

This surprises many returning learners. Dividing by one half asks how many half units fit in 12. There are 24.

Use meaning to check the direction of the result.

Final Number Sense Check

Choose one answer from this chapter and explain why it is reasonable before proving it exactly. Then choose one answer that would be impossible because of sign or size.

This two part check trains the habit you need under pressure: predict the neighborhood, calculate, and compare the result with the prediction.

In a Nutshell

Number sense supports every later chapter.

Understand place value. Estimate the size of an answer before calculating. Choose operations from relationships and units. Follow the order of operations. Treat signs deliberately. Use factors and multiples to prepare for fractions and algebra.

Do not label repeated errors “careless.” Give them a name and a prevention rule.

> ***A reasonable answer is not always correct, but an unreasonable answer is a reason to stop.***

Chapter 7: Connect Fractions, Decimals, and Percentages

Maria could calculate 20 percent of a price at work, but a fraction question on the placement sample stopped her.

She did not think of those skills as related.

At work, 20 percent meant moving a decimal or multiplying by 0.20. In school, $\frac{1}{5}$ belonged to a chapter about common denominators. The ideas had been stored in separate rooms.

Then Maria wrote:

$$\frac{1}{5} = 0.2 = 20\%$$

One quantity. Three forms.

That line did more than help her remember a conversion. It changed the chapter from a collection of rules into a connected system.

The principle is:

> *Fractions, decimals, and percentages are different ways to describe the same quantities. Learn the connections first, then the procedures become easier to remember and check.*

Read a Fraction as Division

A fraction has a numerator on top and a denominator on the bottom.

$$\frac{3}{4}$$

The denominator 4 tells you the whole is divided into four equal parts. The numerator 3 tells you that three parts are being considered.

A fraction also means division:

$$\frac{3}{4} = 3 \div 4 = 0.75$$

This interpretation helps with conversion and with algebraic fractions later.

The denominator cannot be zero because division by zero is undefined.

Simplify Without Changing Value

To simplify a fraction, divide the numerator and denominator by the same nonzero factor.

$$\frac{18}{24}$$

Both numbers are divisible by 6:

$$\frac{18 \div 6}{24 \div 6} = \frac{3}{4}$$

The value has not changed. You have renamed the same quantity.

A fast method is to divide by the greatest common factor. A slower method using several common factors is also valid.

Example:

$$\frac{36}{54}$$

The greatest common factor is 18:

$$\frac{36 \div 18}{54 \div 18} = \frac{2}{3}$$

Check by converting both to decimals if needed:

$$36 \div 54 \approx 0.667$$

$$2 \div 3 \approx 0.667$$

Add and Subtract Fractions

Fractions need a common denominator before addition or subtraction because the pieces must be the same size.

$$\frac{1}{4} + \frac{2}{4} = \frac{3}{4}$$

The denominators are already equal.

For unlike denominators:

$$\frac{2}{3} + \frac{1}{4}$$

The least common denominator is 12.

$$\frac{2}{3} = \frac{8}{12}$$

$$\frac{1}{4} = \frac{3}{12}$$

Then:

$$\frac{8}{12} + \frac{3}{12} = \frac{11}{12}$$

Do not add denominators. The expression

$$\frac{2+1}{3+4} = \frac{3}{7}$$

does not preserve the size of the pieces.

For subtraction:

$$\frac{5}{6} - \frac{1}{4}$$

The least common denominator is 12.

$$\frac{10}{12} - \frac{3}{12} = \frac{7}{12}$$

Estimate before calculating. $\frac{5}{6}$ is a little less than 1, and $\frac{1}{4}$ is 0.25, so the result should be a little more than one half. $\frac{7}{12}$ fits.

Multiply and Divide Fractions

To multiply fractions, multiply numerators and denominators, then simplify.

$$\frac{3}{5} \times \frac{10}{9}$$

You can simplify before multiplying.

The 3 and 9 share a factor of 3:

$$\frac{1}{5} \times \frac{10}{3}$$

The 10 and 5 share a factor of 5:

$$1 \times \frac{2}{3} = \frac{2}{3}$$

To divide by a fraction, multiply by its reciprocal.

$$\frac{3}{4} \div \frac{2}{5}$$

$$\frac{3}{4} \times \frac{5}{2} = \frac{15}{8} = 1\frac{7}{8}$$

Why does the reciprocal appear? Division asks how many groups of the second quantity fit in the first. Multiplying by the reciprocal performs that comparison.

Use the rule, but keep the meaning nearby.

Work With Mixed Numbers

A mixed number combines a whole number and a fraction.

$$2\frac{1}{3}$$

Convert to an improper fraction:

$$2\frac{1}{3} = \frac{2\times3+1}{3} = \frac{7}{3}$$

To convert $\frac{11}{4}$ to a mixed number, divide:

$$11 \div 4 = 2\, remainder\; 3$$

So:

$$\frac{11}{4} = 2\frac{3}{4}$$

For multiplication and division, converting mixed numbers to improper fractions is usually safest.

Understand Decimal Place Value

Align decimal points when adding or subtracting.

$$14.7 + 2.385$$

Write:

$$14.700$$
$$+\ 2.385$$
$$= 17.085$$

For multiplication, multiply as whole numbers, then place the decimal according to the total number of decimal places.

$$2.4 \times 0.35$$

Compute $24 \times 35 = 840$.

There are three decimal places in total, so:

$$2.4 \times 0.35 = 0.840 = 0.84$$

Estimate: 2.4×0.35 is less than $2.4 \times 0.5 = 1.2$. The answer 0.84 is reasonable.

For division by a decimal, move the decimal in both numbers so the divisor becomes a whole number.

$$7.2 \div 0.3$$

Multiply both by 10:

$$72 \div 3 = 24$$

Convert Between Forms

Fraction to decimal:

Divide numerator by denominator.

$$\frac{3}{8} = 3 \div 8 = 0.375$$

Decimal to percent:

Multiply by 100, or move the decimal two places right.

$$0.375 = 37.5\%$$

Percent to decimal:

Divide by 100, or move the decimal two places left.

$$6.5\% = 0.065$$

Percent to fraction:

Write the percent over 100 and simplify.

$$35\% = \frac{35}{100} = \frac{7}{20}$$

Build a small conversion table you can recall quickly:

$$\frac{1}{2} = 0.5 = 50\%$$
$$\frac{1}{4} = 0.25 = 25\%$$
$$\frac{3}{4} = 0.75 = 75\%$$
$$\frac{1}{5} = 0.2 = 20\%$$
$$\frac{1}{10} = 0.1 = 10\%$$

These anchors help you estimate unfamiliar values.

Solve Percent Problems

Three common percent questions are:

Find the part

What is 18 percent of 250?

Convert 18 percent to 0.18:

$$0.18 \times 250 = 45$$

Find the whole

45 is 18 percent of what number?

$$45 = 0.18x$$
$$x = 45 \div 0.18 = 250$$

Find the percent

45 is what percent of 250?

$$\frac{45}{250} = 0.18 = 18\%$$

Use the relationship:

$$part = percent\ as\ a\ decimal \times whole$$

Then solve for the missing quantity.

Percent Increase and Decrease

A price of \$80 increases by 15 percent.

First find the increase:

$$0.15 \times 80 = 12$$

Then add:

$$80 + 12 = 92$$

A faster method is to multiply by 1.15:

$$80 \times 1.15 = 92$$

For a 15 percent decrease, multiply by 0.85:

$$80 \times 0.85 = 68$$

Do not confuse the amount of change with the new total.

A 15 percent discount on $80 is $12. The sale price is $68.

Maria's Error Pattern Log showed that she often stopped after finding the discount. Her prevention rule became:

> *Circle whether the question asks for the change or the final amount.*

Conversion Grid Action

Create a grid with four columns:

Fraction

Decimal

Percent

Meaning or estimate

Fill in at least 12 values, including the five anchors above and these:

$$\frac{1}{8}, \frac{3}{8}, \frac{2}{5}, \frac{3}{5}, \frac{1}{3}, \frac{2}{3}, \frac{1}{20}$$

For repeating decimals such as $\frac{1}{3}$, use an approximation:

$$\frac{1}{3} \approx 0.333 \approx 33.3\%$$

The exact decimal repeats. Your test may ask for a fraction, a rounded decimal, or a percent.

Practice Set

1. Simplify $\frac{28}{42}$.
2. Add $\frac{3}{8} + \frac{1}{6}$.
3. Subtract $\frac{7}{10} - \frac{1}{4}$.

4. Multiply $\frac{5}{12} \times \frac{18}{25}$.
5. Divide $\frac{4}{9} \div \frac{2}{3}$.
6. Convert $2\frac{3}{5}$ to an improper fraction.
7. Convert 0.045 to a percent.
8. Convert 62 percent to a decimal and fraction.
9. Find 35 percent of 240.
10. A $160 fee decreases by 12 percent. What is the new fee?

Answers:

1. $\frac{2}{3}$
2. $\frac{13}{24}$
3. $\frac{9}{20}$
4. $\frac{3}{10}$
5. $\frac{2}{3}$
6. $\frac{13}{5}$
7. 4.5 percent
8. $0.62 = \frac{31}{50}$
9. 84
10. $140.80

For question 10:

$$160 \times 0.88 = 140.80$$

Use One Quantity Three Forms

Choose a fraction and express it in three forms.

Example:

$$\frac{3}{5}$$

Decimal:

$$3 \div 5 = 0.6$$

Percent:

$$0.6 \times 100\% = 60\%$$

Now place it on a number line between 0 and 1. It belongs a little past one half.

Repeat with:

$$\frac{7}{8}, \quad \frac{3}{20}, \quad \frac{11}{25}$$

Answers:

$$\frac{7}{8} = 0.875 = 87.5\%$$

$$\frac{3}{20} = 0.15 = 15\%$$

$$\frac{11}{25} = 0.44 = 44\%$$

This exercise builds magnitude, not only conversion.

Compare Without a Calculator

Compare:

$$\frac{5}{6}$$

and:

$$82\%$$

You can convert 82 percent to 0.82. The fraction $\frac{5}{6}$ is about 0.833, so $\frac{5}{6}$ is larger.

Or compare both with a common benchmark. $\frac{5}{6}$ is 83.3 percent approximately, which exceeds 82 percent.

Now compare:

$$\frac{7}{20}$$

and:

$$0.4$$

$$\frac{7}{20} = 0.35$$

So 0.4 is larger.

Distinguish Percentage Points

A rate rises from 30 percent to 40 percent.

The increase is 10 percentage points.

The percent increase relative to the original rate is:

$$\frac{40-30}{30} \times 100\% = 33.3\% \; approximately$$

These are not the same statement.

Placement questions may use "percent increase," which requires division by the original value. Everyday reports may use "percentage point increase," which subtracts the rates.

Read the wording.

Solve Backward Percent Problems

After a 20 percent discount, a price is $72. Find the original price.

The sale price is 80 percent of the original.

$$0.80x = 72$$
$$x = 90$$

Do not add 20 percent of 72. The 20 percent discount was based on the original amount, not the sale price.

Check:

$$20\% \; of \; 90 = 18$$
$$90 - 18 = 72$$

Backward problems reveal whether you understand the relationship or only one procedure.

Divide Decimals With Meaning

Consider:

$$4.8 \div 0.6$$

Move both decimals one place right:

$$48 \div 6 = 8$$

This works because multiplying both dividend and divisor by 10 does not change the quotient.

Check the meaning:

How many groups of 0.6 fit in 4.8? Eight groups.

Now consider:

$$0.84 \div 2.1$$

Multiply both by 10:

$$8.4 \div 21 = 0.4$$

Check:

$$2.1(0.4) = 0.84$$

Use Fraction Benchmarks

Compare a fraction with 0, one half, and 1.

$$\frac{7}{15}$$

Half of 15 is 7.5, so $\frac{7}{15}$ is slightly less than one half.

$$\frac{11}{12}$$

The numerator is one less than the denominator, so the fraction is close to 1.

Benchmarks help you reject an incorrect decimal conversion.

If $\frac{7}{15}$ is entered as 4.67, the result is impossible because a proper fraction must lie between 0 and 1.

Mixed Number Addition

Add:

$$2\frac{3}{4} + 1\frac{2}{3}$$

Convert to improper fractions:

$$\frac{11}{4} + \frac{5}{3}$$

Common denominator 12:

$$\frac{33}{12} + \frac{20}{12} = \frac{53}{12} = 4\frac{5}{12}$$

You can also add whole and fractional parts separately, then regroup. Use the method that keeps your work clearest.

Percentage Comparison Challenge

A quantity falls from 80 to 68.

Amount of decrease:

$$80 - 68 = 12$$

Percent decrease:

$$\frac{12}{80} \times 100\% = 15\%$$

Now reverse the direction. A rise from 68 to 80 is:

$$\frac{12}{68} \times 100\% \approx 17.6\%$$

The same amount of change produces different percentages because the original bases differ.

Percent change always asks, "Compared with what starting value?"

Conversion Without Symbols

Explain one conversion aloud without using the phrase "move the decimal."

For example: 0.37 means 37 hundredths, so it is 37 percent. This explanation survives when the number is 0.037, which means 37 thousandths and equals 3.7 percent.

Meaning is more reliable than a direction rule remembered under pressure.

Final Practice Rule

When forms change, keep the quantity fixed. A conversion should rename the value, not enlarge or shrink it. Use an estimate to confirm that the fraction, decimal, and percent still describe the same point.

In a Nutshell

Fractions, decimals, and percentages belong to one system.

A fraction means part of a whole and division. Common denominators create equal sized pieces for addition and subtraction. Multiplication and division follow their own rules. Decimals depend on place value. Percent means "per hundred."

Convert among forms until you can see the same quantity wearing three labels.

> *When the forms connect, the rules stop feeling random.*

Chapter 8: Use Ratios, Rates, and Units

A machine produces 420 parts in six hours. At the same rate, how many parts should it produce in nine hours? Before choosing an operation, write the units. They reveal that the task is to find parts per hour and then scale that rate to nine hours.

> ***When a problem involves comparison, rate, proportion, or conversion, write the units first. Units often reveal which quantities should multiply and which should divide.***

Ratios Compare Quantities

A ratio compares two quantities.

If a class has 12 adult learners and 18 recent school graduates, the ratio of adult learners to recent graduates is:

$$12:18$$

or:

$$\frac{12}{18}=\frac{2}{3}$$

Order matters. The ratio of recent graduates to adult learners is:

$$18:12=\frac{3}{2}$$

A ratio may compare quantities with the same units, such as students to students, or different units, such as miles to hours.

Always label the order in words before simplifying.

Rates Include Different Units

A rate compares quantities with different units.

Examples:

$$60\ miles\ per\ hour$$
$$\$18.50\ per\ hour$$
$$12\ credits\ per\ semester$$

A unit rate has a denominator of one unit.

If five notebooks cost $12.50:

$$\$12.50 \div 5=\$2.50$$

The unit rate is $2.50 per notebook.

Unit rates help you compare choices that use different package sizes or time periods.

Priya compared two study platforms. One cost \$36 for three months. The other cost \$70 for six months.

First platform:

$$36 \div 3 = 12$$

\$12 per month.

Second platform:

$$70 \div 6 \approx 11.67$$

About \$11.67 per month.

The second had the lower monthly rate, though price was not the only decision factor.

Proportions State Equal Ratios

A proportion is an equation showing that two ratios are equal.

$$\frac{3}{5} = \frac{x}{20}$$

You can solve by understanding scale. Since 5 is multiplied by 4 to reach 20, multiply 3 by 4:

$$x = 12$$

You can also cross multiply:

$$3 \times 20 = 5x$$
$$60 = 5x$$
$$x = 12$$

Cross multiplication works because equal fractions remain equal when both sides are multiplied by the denominators. It is not a universal instruction for every pair of fractions on a page. Use it when you have a true proportion.

Test Whether a Relationship Is Proportional

A proportional relationship has a constant ratio.

Suppose a table shows:

$$2\ hours \rightarrow \$36$$
$$5\ hours \rightarrow \$90$$
$$8\ hours \rightarrow \$144$$

Divide pay by hours:

$$36 \div 2 = 18$$
$$90 \div 5 = 18$$
$$144 \div 8 = 18$$

The rate is always \$18 per hour, so the relationship is proportional.

Now suppose a service charges a \$20 starting fee plus \$18 per hour.

Two hours cost \$56. Five hours cost \$110.

$$56 \div 2 = 28$$

$$110 \div 5 = 22$$

The ratio is not constant. The fixed fee makes the relationship nonproportional.

This distinction matters in tables, graphs, formulas, and word problems.

Convert Units With Conversion Factors

A conversion factor is a ratio equal to one because the numerator and denominator represent the same quantity.

For example:

$$\frac{12\ inches}{1\ foot}$$

To convert 7 feet to inches:

$$7\ feet \times \frac{12\ inches}{1\ foot} = 84\ inches$$

The feet units cancel, leaving inches.

To convert 90 minutes to hours:

$$90\ minutes \times \frac{1\ hour}{60\ minutes} = 1.5\ hours$$

Choose the conversion factor so the unwanted unit cancels.

This method is sometimes called dimensional analysis. The name sounds advanced. The action is simple: arrange units so only the desired unit remains.

Use Multiple Conversions

Convert 55 miles per hour to feet per second.

Use:

$$1\ mile = 5{,}280\ feet$$

$$1\ hour = 3{,}600\ seconds$$

Then:

$$55\frac{miles}{hour} \times \frac{5{,}280\ feet}{1\ mile} \times \frac{1\ hour}{3{,}600\ seconds}$$

Miles and hours cancel:

$$55 \times \frac{5{,}280}{3{,}600} \approx 80.7$$

The rate is about 80.7 feet per second.

You may not need this exact conversion, but the structure appears in technical, science, health, and placement contexts.

Scale Drawings and Maps

A scale relates a drawing to the real object.

Suppose a floor plan uses:

$$1\ inch = 4\ feet$$

A wall measures 3.5 inches on the plan.

$$3.5 \times 4 = 14$$

The wall is 14 feet long.

Keep drawing units and real units labeled. A common Error Pattern is to reverse the scale.

Carlos wrote this prevention rule:

> *Read the scale aloud: one drawing inch represents four real feet.*

Rates in Adult Contexts

Placement problems often use ordinary settings. The mathematics is the same across contexts.

Work rate

A worker completes 18 applications in three hours.

$$18 \div 3 = 6$$

Six applications per hour.

Travel rate

A vehicle travels 156 miles in three hours.

$$156 \div 3 = 52$$

52 miles per hour.

Price rate

Eight pounds of material cost $22.40.

$$22.40 \div 8 = 2.80$$

$2.80 per pound.

Credit load

A program requires 60 credits over four terms.

$$60 \div 4 = 15$$

An average of 15 credits per term.

The unit rate is a translation tool. It turns a total into a reusable relationship.

Do Not Use Proportions Automatically

Some problems look like ratio problems but include fixed amounts, changing rates, or percentage changes.

Example:

A taxi charge is $4 plus $2.50 per mile.

The cost for five miles is:

$$4 + 2.50(5) = 16.50$$

You cannot set up a direct proportion between miles and total cost because the $4 fee exists even at zero miles.

Example:

A salary increases by 5 percent each year. The second increase is based on the new salary, not the original amount. This is repeated percentage growth, not one simple proportion across all years.

Ask whether the ratio remains constant.

The Units First Method

Use this four step method.

Name the requested unit.

What must the answer be measured in?

Write every given quantity with units.

Do not write 420 and 6. Write 420 parts and six hours.

Arrange multiplication and division so unwanted units cancel.

If units do not cancel as expected, the setup may be wrong.

Check the direction.

Should the answer be larger or smaller? Does the rate make sense?

Apply it:

A student reads 84 pages in six hours. At the same rate, how many pages in 2.5 hours?

Requested unit: pages.

Unit rate:

$$84\, pages \div 6\, hours = 14\, pages\ per\ hour$$

Then:

$$14\, pages\ per\ hour \times 2.5\, hours = 35\, pages$$

Answer: 35 pages.

Practice Set

1. Simplify the ratio 24 to 36.
2. A car travels 285 miles on 9.5 gallons. Find miles per gallon.
3. Six shirts cost $87. Find the unit price.
4. Solve $\frac{7}{12} = \frac{x}{48}$.
5. A recipe uses three cups of flour for 24 servings. How many cups are needed for 40 servings?
6. Convert 6.5 feet to inches.
7. Convert 150 minutes to hours.
8. A map scale is one inch to 18 miles. Two cities are 3.25 inches apart. Find the real distance.
9. Determine whether the table is proportional: 2 items cost $7, 4 items cost $14, and 7 items cost $24.50.
10. A service charges $30 plus $12 per hour. Find the cost for 4.5 hours.

Answers:

1. $2:3$
2. 30 miles per gallon
3. $14.50 per shirt
4. $x = 28$
5. Five cups
6. 78 inches
7. 2.5 hours
8. 58.5 miles
9. Yes. Each item costs $3.50.
10. $84

For question 5:

$$\frac{3}{24} = \frac{x}{40}$$
$$24x = 120$$
$$x = 5$$

For question 10:

$$30 + 12(4.5) = 30 + 54 = 84$$

Distinguish Direct and Inverse Change

In a direct proportional relationship, increasing one quantity by a factor increases the other by the same factor.

If three notebooks cost \$12, six notebooks cost \$24 at the same unit price.

The number of notebooks doubled. The cost doubled.

Some relationships are inverse.

If one worker needs 12 hours for a task and two equally productive workers divide the work perfectly, the time may fall to six hours. More workers, less time.

Do not force an inverse situation into a direct proportion.

Ask:

When one quantity increases, should the other increase or decrease?

Then ask whether the problem assumes a constant rate and perfect sharing. Real situations may include limits, setup time, or changing efficiency.

Check a Proportion by Units

Suppose four pounds cost \$11. What do seven pounds cost?

A useful proportion is:

$$\frac{\$11}{4\ pounds} = \frac{x\ dollars}{7\ pounds}$$

Dollars appear in the numerator on both sides. Pounds appear in the denominator.

An inconsistent setup such as:

$$\frac{\$11}{4\ pounds} = \frac{7\ pounds}{x\ dollars}$$

mixes the order.

Cross multiplication can still produce a number, but the setup does not represent equal rates.

Units can catch the error before calculation.

Use Rate Language Precisely

"A per B" usually means:

$$\frac{A}{B}$$

Miles per hour:

$$\frac{miles}{hour}$$

Dollars per item:

$$\frac{dollars}{item}$$

Students per instructor:

$$\frac{students}{instructor}$$

When the desired answer is hours, arrange the calculation so hours remain.

If a machine processes 15 items per hour and must process 90 items:

$$90\ items \times \frac{1\ hour}{15\ items} = 6\ hours$$

The reciprocal of the rate appears because items must cancel.

Rate Repair Challenge

Find one rate from your ordinary week.

Examples include travel, pay, reading, data use, or cost per unit.

Record the total quantities, calculate the unit rate, and write one sentence interpreting it.

Then change one input and predict the output if the rate remains constant.

This turns a formula into a relationship you can test.

Use Proportion Estimates

Before solving:

$$\frac{4}{7} = \frac{x}{35}$$

Notice that 35 is five times 7. The answer should be five times 4, or 20.

This scale reasoning is faster and safer than cross multiplication when the relationship is obvious.

For:

$$\frac{7}{12} = \frac{x}{50}$$

The scale is not as clean. Estimate first.

$$\frac{7}{12} \approx 0.583$$

A little more than half of 50 is a little more than 25. The exact answer should be about 29.

Cross multiply:

$$12x = 350$$
$$x \approx 29.17$$

The estimate confirms the result.

Keep Rates Consistent

Do not compare miles per hour with miles per minute until the time units match.

A rate of 1.2 miles per minute equals:

$$1.2 \times 60 = 72\ miles\ per\ hour$$

A rate of 65 miles per hour is therefore lower.

Convert before comparing.

This applies to prices, production, data, and credit loads. A weekly cost and monthly cost cannot be compared directly until they use the same period.

Ratio Language Check

Write the order of every ratio in words before simplifying it.

If a room has 12 adults and eight children, adults to children is $12:8$, while children to adults is $8:12$. Both simplify, but they answer different questions.

The order stated in the sentence controls the order in the ratio.

In a Nutshell

Ratios compare quantities. Rates compare quantities with different units. Unit rates reduce the comparison to one. Proportions state that two ratios are equal.

Write units beside every number. Use conversion factors so unwanted units cancel. Check whether a relationship is truly proportional before using cross multiplication.

> ***Units are not labels added after the work. They are instructions for the work.***

Chapter 9: Translate Algebra First

Mei could follow a worked algebra example line by line. The moment the numbers changed, she stopped.

The problem was not always the calculation. It was the language.

A coefficient sounded like a piece of equipment. A term sounded like a school semester. "Evaluate the expression" sounded as if she was being asked for an opinion.

Once the words became ordinary, the symbols became less threatening.

The expression:

$$3x + 5$$

means three times a number, plus five.

If $x = 4$, the expression means:

$$3(4) + 5 = 17$$

The chapter principle is:

> *Algebra becomes easier when you translate before you manipulate. Name what each symbol means, identify the structure, and only then choose a procedure.*

Variables Represent Quantities

A variable is a symbol used to represent a number or quantity.

In:

$$x + 7$$

x may represent an unknown number or a quantity that can change.

In a formula such as:

$$d = rt$$

d represents distance, r represents rate, and t represents time.

A variable is not a special kind of number. It stands in the place where a number can go.

Learn the Parts of an Expression

Consider:

$$4x - 3y + 8$$

The terms are:

$$4x, \quad -3y, \quad 8$$

Terms are separated by addition or subtraction signs.

The coefficient of x is 4.

The coefficient of y is -3.

The constant is 8 because it does not contain a variable.

This vocabulary helps you read instructions such as "combine like terms" or "identify the coefficient."

Translate Words Into Symbols

Common translations include:

sum means addition
difference means subtraction
product means multiplication
quotient means division
more than means addition
less than reverses subtraction order
twice means multiply by two
of often means multiplication
per often creates a rate or division

Examples:

Five more than a number:

$$x + 5$$

Five less than a number:

$$x - 5$$

Five less than twice a number:

$$2x - 5$$

The difference between five and a number:

$$5 - x$$

A number divided by seven:

$$\frac{x}{7}$$

Seven divided by a number:

$$\frac{7}{x}$$

Order matters.

Tasha translated "six less than a number" as $6 - x$. Her Error Pattern came from reading left to right instead of following meaning. Her prevention rule was:

Replace "less than" with "subtracted from."

Six less than a number becomes six subtracted from the number:

$$x - 6$$

Evaluate by Substitution

To evaluate an expression, replace each variable with its given value and simplify.

Evaluate:

$$2x^2 - 3x + 1$$

when $x = -2$.

Substitute with parentheses:

$$2(-2)^2 - 3(-2) + 1$$

Calculate the exponent:

$$(-2)^2 = 4$$

Then:

$$2(4) + 6 + 1 = 15$$

Parentheses protect the negative value. Without them, -2^2 can be misread as $-\left(2^2\right) = -4$.

Use parentheses every time you substitute a negative number.

Combine Like Terms

Like terms have the same variable part.

$$3x + 5x = 8x$$

$$7y^2 - 2y^2 = 5y^2$$

You cannot combine:

$$3x + 5$$

because one term contains x and the other does not.

You also cannot combine:

$$2x + 3x^2$$

because x and x^2 are different variable parts.

Think of the coefficient as the count.

Three x units plus five x units equals eight x units.

Use the Distributive Property

The distributive property connects multiplication and addition.

$$a(b + c) = ab + ac$$

Example:

$$3(x + 4) = 3x + 12$$

Multiply 3 by every term inside the parentheses.

With subtraction:

$$-2(3x - 5)$$
$$= -6x + 10$$

The negative multiplier affects both terms.

A common Error Pattern is distributing to the first term only.

Incorrect:

$$4(x + 3) = 4x + 3$$

Correct:

$$4(x + 3) = 4x + 12$$

Renee drew two arrows from the outside multiplier, one to each term. After the habit stabilized, she no longer needed the arrows.

Simplify Expressions in Stages

Simplify:

$$3(2x - 4) + 5x + 7$$

Distribute:

$$6x - 12 + 5x + 7$$

Combine like terms:

$$11x - 5$$

You cannot solve for x because there is no equation. The instruction is to simplify, not solve.

This distinction appears often:

An **expression** does not contain an equals sign.

An **equation** states that two expressions are equal.

You simplify an expression. You solve an equation.

Recognize Equivalent Expressions

Two expressions are equivalent if they produce the same value for every permitted value of the variable.

$$2(x + 3)$$

and:

$$2x + 6$$

are equivalent.

You can verify by distribution or by testing several values. Testing values supports the idea but does not prove equivalence for every possible number. Algebraic properties provide the proof.

Placement tests may ask which expression is equivalent to another. Distribute, factor, or combine like terms rather than relying on appearance.

Use Formulas as Compressed Relationships

A formula summarizes a relationship.

The area of a rectangle is:

$$A = lw$$

If length is 8 and width is 5:

$$A = 8(5) = 40$$

The formula does not need to be mysterious. Translate it:

Area equals length times width.

The simple interest formula is:

$$I = Prt$$

Translate:

Interest equals principal times rate times time.

In a placement context, use the formula exactly as given and pay attention to units. This book uses financial examples only to teach mathematics, not to provide financial advice.

Translation Practice

Write an algebraic expression for each phrase.

1. Nine more than a number n
2. Four times a number x
3. Seven less than three times y
4. The quotient of p and five
5. Five subtracted from twice m
6. The total cost of x tickets at \$12 each plus a \$4 fee
7. A number k decreased by 18 percent of itself

Answers:

1. $n + 9$
2. $4x$
3. $3y - 7$

4. $\frac{p}{5}$
5. $2m - 5$
6. $12x + 4$
7. $k - 0.18k$, which simplifies to $0.82k$

Simplification Practice

1. $5x + 2x - 4$
2. $3y + 8 - y + 2$
3. $4(a + 6)$
4. $-3(2m - 5)$
5. $2(3x + 1) - 4x$
6. Evaluate $3p - 7$ when $p = 5$.
7. Evaluate $2q^2 + q$ when $q = -3$.
8. Simplify $5(2x - 1) + 3(x + 4)$.

Answers:

1. $7x - 4$
2. $2y + 10$
3. $4a + 24$
4. $-6m + 15$
5. $2x + 2$
6. 8
7. 15
8. $13x + 7$

Check question 7:

$$2(-3)^2 + (-3) = 2(9) - 3 = 15$$

The Translate Mark Solve Habit

For every algebra question:

Translate

Say what the expression or equation means in words.

Mark

Circle grouping symbols, negative signs, exponents, and the requested task.

Solve or simplify

Apply the appropriate property one line at a time.

This habit prevents a common problem: performing a correct procedure on the wrong structure.

Eric saw $3x^2$ and squared both 3 and x. The expression means $3 \times x^2$, not $(3x)^2$. Marking the exponent showed exactly what was squared.

Distinguish Similar Looking Expressions

Compare:

$$3x^2$$

and:

$$(3x)^2$$

The first means:

$$3 \times x^2$$

The second means:

$$9x^2$$

Compare:

$$-2^2$$

and:

$$(-2)^2$$

The first is:

$$-\left(2^2\right) = -4$$

The second is:

$$4$$

Grouping symbols control meaning.

Now compare:

$$\frac{x+3}{5}$$

and:

$$x + \frac{3}{5}$$

In the first expression, the entire sum is divided by five. In the second, only 3 is divided by five.

Copying a structure incorrectly creates a different problem.

Translate a Cost Rule

A service charges a $28 start fee and $16 per hour.

Let h represent hours and C represent total cost.

$$C = 16h + 28$$

Translate each part.

$16h$ is the hourly charge.

28 is the fixed charge.

If the total is $108:

$$108 = 16h + 28$$

The same expression has become an equation because the total is known.

If two services are compared, each gets its own expression.

This is how algebra grows from a description rather than appearing as symbols without a source.

Factor as a Translation

The expression:

$$6x + 18$$

can be factored:

$$6(x + 3)$$

The first form shows two terms.

The second shows six groups of $x + 3$.

Both forms are equivalent. One may be more useful for addition. The other may be more useful for solving or recognizing structure.

Factoring is not limited to the Higher-Math chapter. The basic idea is identifying a shared multiplier.

Create Your Own Expression

Write one ordinary rule with:

a variable amount

a fixed amount

a total

Example:

A course platform charges \$9 per month plus a \$15 activation fee.

$$T = 9m + 15$$

Choose $m = 6$ and evaluate.

$$T = 9(6) + 15 = 69$$

Then reverse the problem. If the total is \$96, solve for m.

$$96 = 9m + 15$$

$$81 = 9m$$

$$m = 9$$

You have translated, evaluated, and solved one relationship.

Use Substitution to Check Equivalence

Are these expressions equivalent?

$$3(x + 2)$$

and:

$$3x + 2$$

Distribution says the first becomes:

$$3x + 6$$

So they are not equivalent.

A quick test value also reveals the difference. Let $x = 1$.

First expression:

$$3(1 + 2) = 9$$

Second expression:

$$3(1) + 2 = 5$$

One counterexample proves they are not equivalent.

Now test:

$$4(x - 3)$$

and:

$$4x - 12$$

Both give the same result because distribution connects them.

Translate Inequality Language

At least means greater than or equal to.

$$x \geq 10$$

At most means less than or equal to.

$$x \leq 10$$

More than means greater than.

$$x > 10$$

Fewer than means less than.

$$x < 10$$

Example:

A program requires at least 60 credits.

$$c \geq 60$$

The endpoint 60 is included.

Example:

A room holds fewer than 40 people.

$$p < 40$$

The endpoint 40 is not included.

Watch Implied Multiplication

Algebra often omits the multiplication sign.

$$5x$$

means:

$$5 \times x$$

$$2(a + b)$$

means:

$$2 \times (a + b)$$

Adjacent parentheses can also mean multiplication:

$$(x + 2)(x - 3)$$

Reading implied multiplication aloud can prevent you from treating symbols as one unexplained block.

Name the Requested Task

Before working, circle the instruction.

Simplify asks for an equivalent expression.

Evaluate asks for a number after values are substituted.

Solve asks for value or values that make an equation true.

Factor asks for a product of simpler expressions.

A correct method for the wrong instruction still produces the wrong response.

Final Translation Rule

Read every algebraic line in words once. If the spoken meaning changes between lines, inspect the operation. Equivalent steps may look different, but they must continue to describe the same quantity or relationship.

In a Nutshell

Algebra is a language for quantities and relationships.

Variables stand for numbers. Terms are separated by addition and subtraction. Coefficients multiply variable parts. Like terms can combine. The distributive property multiplies every term inside a group. Substitution replaces variables with values.

Translate first. Mark the structure. Then calculate.

> *A symbol you can name is a symbol you can work with.*

Chapter 10: Solve Equations With Meaning

An equation is a balance, not a collection of terms to "move." Each step must preserve equality, and every result should be checked in the original equation.

> ***Do not move terms by magic. Preserve equality by performing the same valid operation on both sides, then check the result in the original equation.***

Understand Equality

The equals sign does not mean "the answer comes next."

It means the expression on the left has the same value as the expression on the right.

In:

$$3 + 5 = 10 - 2$$

both sides equal 8.

In:

$$x + 4 = 11$$

the equation asks which value of x makes the two sides equal.

The answer is 7 because:

$$7 + 4 = 11$$

Use Inverse Operations

Inverse operations undo each other.

Addition and subtraction are inverses.

Multiplication and division are inverses.

Squaring and taking a square root can be inverses under appropriate conditions.

Solve:

$$x + 9 = 17$$

Subtract 9 from both sides:

$$x + 9 - 9 = 17 - 9$$

$$x = 8$$

Solve:

$$5x = 35$$

Divide both sides by 5:

$$\frac{5x}{5} = \frac{35}{5}$$

$$x = 7$$

Write the operation on both sides during practice. It makes the equality visible.

Solve Multistep Equations

Solve:

$$3x + 8 = 29$$

Subtract 8 from both sides:

$$3x = 21$$

Divide both sides by 3:

$$x = 7$$

Check:

$$3(7) + 8 = 21 + 8 = 29$$

Correct.

Solve:

$$4(x - 2) + 3 = 19$$

Distribute:

$$4x - 8 + 3 = 19$$

Combine like terms:

$$4x - 5 = 19$$

Add 5:

$$4x = 24$$

Divide by 4:

$$x = 6$$

Check in the original equation:

$$4(6 - 2) + 3 = 4(4) + 3 = 19$$

Checking the original catches errors introduced before simplification.

Work With Variables on Both Sides

Solve:

$$5x + 4 = 2x + 19$$

Subtract $2x$ from both sides:

$$3x + 4 = 19$$

Subtract 4:

$$3x = 15$$

Divide by 3:

$$x = 5$$

You could instead subtract $5x$ from both sides. That route produces negative coefficients but remains valid.

Choose a route that keeps the work clear.

Solve:

$$7 - 2x = x + 16$$

Subtract x:

$$7 - 3x = 16$$

Subtract 7:

$$-3x = 9$$

Divide by -3:

$$x = -3$$

Check:

Left side:

$$7 - 2(-3) = 7 + 6 = 13$$

Right side:

$$-3 + 16 = 13$$

Equations With Fractions

Solve:

$$\frac{x}{5} + 2 = 9$$

Subtract 2:

$$\frac{x}{5} = 7$$

Multiply both sides by 5:

$$x = 35$$

For equations with several fractions, multiply every term by the least common denominator.

Solve:

$$\frac{x}{3} + \frac{1}{2} = \frac{5}{6}$$

The least common denominator is 6.

Multiply every term by 6:

$$6\left(\frac{x}{3}\right) + 6\left(\frac{1}{2}\right) = 6\left(\frac{5}{6}\right)$$

$$2x + 3 = 5$$

$$2x = 2$$

$$x = 1$$

Check:

$$\frac{1}{3} + \frac{1}{2} = \frac{2}{6} + \frac{3}{6} = \frac{5}{6}$$

Equations With Decimals

Solve:

$$0.4x + 1.2 = 5.6$$

Subtract 1.2:

$$0.4x = 4.4$$

Divide by 0.4:

$$x = 11$$

You can also multiply the entire equation by 10 first:

$$4x + 12 = 56$$

Then solve:

$$4x = 44$$

$$x = 11$$

Clearing decimals can reduce place value mistakes.

Understand Special Results

Some equations have no solution.

$$2(x + 3) = 2x + 9$$

Distribute:

$$2x + 6 = 2x + 9$$

Subtract $2x$:

$$6 = 9$$

That statement is false, so no value of x satisfies the equation.

Some equations have infinitely many solutions.

$$3(x + 2) = 3x + 6$$

Distribute:

$$3x + 6 = 3x + 6$$

Subtract $3x$:

$$6 = 6$$

That statement is always true, so every real value of x satisfies the equation.

Do not force a number when the variable disappears.

Solve Inequalities

An inequality compares values.

Symbols include:

$$<, >, \leq, \geq$$

Solve:

$$x + 5 < 12$$

Subtract 5:

$$x < 7$$

The solution includes every number less than 7.

Solve:

$$-3x \geq 12$$

Divide by -3. When multiplying or dividing both sides of an inequality by a negative number, reverse the inequality sign:

$$x \leq -4$$

Why reverse it?

Start with a true statement:

$$2 < 5$$

Multiply by -1:

$$-2 > -5$$

The order reverses on the number line.

Graph Inequality Solutions

For $x < 7$, use an open circle at 7 and shade left.

For $x \leq 7$, use a closed circle at 7 and shade left.

Open means the endpoint is not included.

Closed means it is included.

For $x > -2$, shade right.

The inequality symbol tells you the direction after the variable is isolated.

Rearrange Formulas

Sometimes the variable you need is not already alone.

Start with:

$$d = rt$$

Solve for t.

Divide both sides by r:

$$t = \frac{d}{r}, \quad r \neq 0$$

Start with:

$$A = \frac{1}{2}bh$$

Solve for h.

Multiply both sides by 2:

$$2A = bh$$

Divide by b:

$$h = \frac{2A}{b}, \quad b \neq 0$$

Treat the formula like an equation. Perform the same operation on both sides.

Use the Solve Check Explain Routine

For each equation:

Solve

Show one valid transformation per line.

Check

Substitute the answer into the original equation.

Explain

Name the inverse operation or property used.

Example:

$$2x - 7 = 13$$

Add 7:

$$2x = 20$$

Divide by 2:

$$x = 10$$

Check:

$$2(10) - 7 = 13$$

Explain: Addition undid subtraction, and division undid multiplication.

Practice Set

1. $x - 12 = 9$
2. $6x = 54$
3. $4x + 7 = 31$
4. $3(x + 5) = 27$
5. $7x - 4 = 3x + 20$

6. $\frac{x}{8} - 2 = 5$
7. $0.25x + 3 = 8$
8. $\frac{x}{4} + \frac{1}{3} = \frac{5}{6}$
9. $5 - 2x \leq 13$
10. Solve $P = 2l + 2w$ for w.

Answers:

1. $x = 21$
2. $x = 9$
3. $x = 6$
4. $x = 4$
5. $x = 6$
6. $x = 56$
7. $x = 20$
8. $x = 2$
9. $x \geq -4$
10. $w = \frac{P-2l}{2}$, or $w = \frac{P}{2} - l$

For question 8, multiply every term by 12:

$$3x + 4 = 10$$
$$3x = 6$$
$$x = 2$$

For question 9:

$$5 - 2x \leq 13$$
$$-2x \leq 8$$

Divide by -2 and reverse the sign:

$$x \geq -4$$

Error Check

Review your answers and classify every miss.

Was the error caused by:

a sign

distribution

combining like terms

an inverse operation

fraction arithmetic

decimal arithmetic

reversing an inequality
checking the simplified equation instead of the original
not recognizing a special result
Write a prevention rule.
"Practice equations" is not a prevention rule.
"Write the same operation under both sides" is.

Choose the Clearest First Move

Some equations allow more than one valid first step.

Solve:

$$2(x + 5) = 18$$

Route one distributes:

$$2x + 10 = 18$$
$$2x = 8$$
$$x = 4$$

Route two divides by 2 first:

$$x + 5 = 9$$
$$x = 4$$

Both are valid. The second is shorter.

Now solve:

$$3(x - 2) + 5 = 2x + 11$$

Distribution creates a clearer path:

$$3x - 6 + 5 = 2x + 11$$
$$3x - 1 = 2x + 11$$
$$x = 12$$

Do not search for one sacred order. Search for a valid step that simplifies the structure.

Check Inequalities With a Test Value

Solve:

$$2x + 3 < 11$$
$$2x < 8$$
$$x < 4$$

Choose a value that should work, such as $x = 0$.

$$2(0) + 3 < 11$$
$$3 < 11$$

True.

Choose a value that should not work, such as $x = 5$.

$$13 < 11$$

False.

The test does not replace algebra. It checks direction.

This is especially useful after dividing by a negative number.

Translate Formula Rearrangement

Start with:

$$C = \frac{5}{9}(F - 32)$$

Suppose you need F.

Multiply both sides by $\frac{9}{5}$:

$$\frac{9}{5}C = F - 32$$

Add 32:

$$F = \frac{9}{5}C + 32$$

You do not need to memorize every rearranged form if you can preserve equality.

Explain One Equation Aloud

Choose a missed equation and explain every line without saying "move."

Use phrases such as:

"I subtract 7 from both sides."

"I divide both sides by negative 3."

"I distribute 4 to both terms."

"I combine like terms."

If you cannot name the operation, the step may be memorized without meaning.

> ***What you can explain is easier to rebuild under pressure.***

Solve Literal Equations One Layer at a Time

A literal equation contains several variables.

Solve:

$$V = lwh$$

for h.

Divide both sides by lw:

$$h = \frac{V}{lw}, \quad lw \neq 0$$

The same balance rule applies.

Solve:

$$y = mx + b$$

for x.

Subtract b:

$$y - b = mx$$

Divide by m:

$$x = \frac{y-b}{m}, \quad m \neq 0$$

Literal equations can look harder because several letters remain. The goal is not to produce a number. It is to isolate the requested variable.

Clear Fractions Carefully

Solve:

$$\frac{x-2}{4} + \frac{x}{6} = 3$$

The least common denominator is 12.

Multiply every term by 12:

$$3(x - 2) + 2x = 36$$

Distribute:

$$3x - 6 + 2x = 36$$

$$5x = 42$$

$$x = \frac{42}{5} = 8.4$$

Check in the original equation.

Do not multiply only the denominators or only the first term. The multiplier applies to every term on both sides.

Identify Extraneous Results

Some advanced equations can produce a candidate answer that does not satisfy the original, especially after squaring both sides or working with rational expressions.

Even in basic algebra, checking builds the habit that protects against those later cases.

A candidate is not a solution until the original equation accepts it.

In a Nutshell

An equation is a statement of equality. Preserve that equality by performing the same valid operation on both sides.

Use inverse operations, distribute when needed, combine like terms, and isolate the variable. When dividing an inequality by a negative number, reverse the sign. Rearrange formulas with the same balance method.

Always check in the original.

> ***A solution is not finished when the variable is alone. It is finished when the original equation agrees.***

Chapter 11: Read Graphs, Tables, and Data

Heather understood the numbers in a table but missed the question because she read the wrong direction.

The table listed hours studied in the left column and practice scores in the right. The question asked for the score associated with six hours. Heather located a score of six and read across to the hours column.

Nothing was wrong with her arithmetic. She had not identified the input and output.

The next time, she wrote:

Input: hours Output: score

Then she traced from six hours across the correct row.

The chapter principle is:

> *A table, graph, formula, and written description may show the same relationship in different forms. Identify the input, output, scale, and rate before calculating.*

Use the Coordinate Plane

The coordinate plane has a horizontal x axis and a vertical y axis.

A point is written as an ordered pair:

$$(x, y)$$

The first number tells you how far to move horizontally. The second tells you how far to move vertically.

For $(3, -2)$:

Move three units right and two units down.

The axes divide the plane into four quadrants.

Quadrant I: positive x, positive y

Quadrant II: negative x, positive y

Quadrant III: negative x, negative y

Quadrant IV: positive x, negative y

A point on an axis is not in a quadrant.

Read Scale Before Shape

Graphs can mislead when the scale does not begin at zero or changes by an unexpected amount.

Before interpreting a graph, check:

the title
the labels on both axes
the unit on each axis
the starting value
the interval between marks
whether the scale is consistent

Suppose a vertical axis begins at 94 rather than zero. A change from 96 to 98 may look visually dramatic. The actual increase is two units.

Do not describe a graph as "shooting up" until you calculate the change.

Understand Slope as Rate of Change

Slope measures the change in y compared with the change in x.

$$m = \frac{y_2 - y_1}{x_2 - x_1}, \quad x_2 \neq x_1$$

This is often described as rise over run.

Find the slope between $(2, 5)$ and $(6, 13)$.

$$m = \frac{13-5}{6-2}$$

$$m = \frac{8}{4} = 2$$

The output increases by two units for each one unit increase in the input.

If x represents hours and y represents dollars, the slope may represent dollars per hour.

Units give slope meaning.

Recognize Positive Negative Zero Undefined

A positive slope rises from left to right.

A negative slope falls from left to right.

A horizontal line has slope zero because y does not change.

A vertical line has undefined slope because the change in x is zero and division by zero is undefined.

Do not confuse "undefined" with zero.

Use Slope Intercept Form

A linear equation can be written:

$$y = mx + b$$

where:

m is slope.
b is the y intercept, the value of y when $x = 0$.
Example:

$$y = 18x + 25$$

If x is the number of hours and y is total cost, the slope 18 means \$18 per hour. The intercept 25 means an initial fee of \$25.

This is not a proportional relationship because the graph does not pass through the origin.

For a proportional relationship:

$$y = kx$$

The intercept is zero, and k is the constant of proportionality.

Build a Table From a Formula

Use:

$$y = 2x + 3$$

Choose values of x:

$$x = 0, 1, 2, 3$$

Calculate y:

When $x = 0$:

$$y = 2(0) + 3 = 3$$

When $x = 1$:

$$y = 5$$

When $x = 2$:

$$y = 7$$

When $x = 3$:

$$y = 9$$

The ordered pairs are:

$$(0, 3), (1, 5), (2, 7), (3, 9)$$

The output rises by two each time the input rises by one.

Understand Functions

A function assigns exactly one output to each permitted input.

A person can have one current age for a given date. That relationship can act like a function.

A single input cannot produce two different outputs in the same function rule.

In a table, if $x = 4$ appears with both $y = 7$ and $y = 10$, the relation is not a function.

On a graph, the vertical line test checks this. If any vertical line crosses the graph more than once, the graph is not a function.

Function notation:

$$f(x) = 3x - 2$$

Find $f(5)$:

$$f(5) = 3(5) - 2 = 13$$

The notation $f(5)$ does not mean $f \times 5$. It means the output of the function when the input is 5.

Read Tables for Patterns

Consider:

Hours	Total Cost
0	30
1	42
2	54
3	66

The cost increases by $12 per hour. The starting cost is $30.

Formula:

$$C = 12h + 30$$

The table, formula, and graph represent the same relationship.

Ask four questions:

1. What is the input?
2. What is the output?
3. How much does the output change?
4. What is the output when the input is zero?

These questions often reveal slope and intercept without formal calculation.

Use Mean Median Mode and Range

The mean is the sum divided by the number of values.

Data:

$$4, 7, 7, 9, 13$$

Mean:

$$\frac{4+7+7+9+13}{5} = \frac{40}{5} = 8$$

The median is the middle value after ordering:

$$7$$

The mode is the most frequent value:

$$7$$

The range is maximum minus minimum:

$$13 - 4 = 9$$

A data set can have no mode, one mode, or more than one mode.

Know When Mean Can Mislead

Consider monthly incomes:

$$2,400,\ 2,500,\ 2,600,\ 2,700,\ 15,000$$

Mean:

$$\frac{25,200}{5} = 5,040$$

Median:

$$2,600$$

The high value pulls the mean upward. In this skewed data set, the median better represents a typical monthly income than the mean does.

This does not make the mean wrong. It means the two measures answer different questions.

A placement problem may ask which measure best represents typical data. Look for outliers and skew.

Interpret Percent Change in Data

Suppose enrollment rises from 800 to 920.

The amount of increase is:

$$920 - 800 = 120$$

The percent increase is based on the original value:

$$\frac{120}{800} \times 100\% = 15\%$$

A common error is dividing by the new value. Use the original as the comparison base unless the problem defines another base.

Graph and Data Practice

1. Identify the quadrant of $(-4, 6)$.
2. Find the slope between $(1, 3)$ and $(5, 15)$.

3. What is the slope of $y = -4x + 7$?
4. What is the y intercept of $y = 6x - 9$?
5. Find $f(-2)$ if $f(x) = x^2 + 3x$.
6. Determine whether the relation $\{(1,4), (2,5), (1,7)\}$ is a function.
7. Find the mean of $6, 8, 8, 10, 13$.
8. Find the median and range of $3, 5, 9, 11, 21, 30$.
9. A value rises from 250 to 290. Find the percent increase.
10. A service follows $C = 15h + 40$. Explain the meaning of 15 and 40.

Answers:

1. Quadrant II
2. 3
3. -4
4. -9
5. -2
6. No. Input 1 has two outputs.
7. 9
8. Median 10, range 27
9. 16 percent
10. The service costs $15 per hour plus a $40 starting charge.

For question 5:

$$f(-2) = (-2)^2 + 3(-2) = 4 - 6 = -2$$

For question 9:

$$\frac{290-250}{250} \times 100\% = \frac{40}{250} \times 100\% = 16\%$$

The Four Form Action

Choose one simple real situation, such as hourly pay plus a fixed bonus, a phone plan, or pages read over time.

Represent it in four forms:

1. A sentence
2. A table
3. A formula
4. A sketch of a graph

Then explain what the slope or rate means and what the starting value means.

This exercise forces translation across forms. Placement questions often test exactly that movement.

Match the Story to the Graph

Consider three situations.

A savings account starts at \$200 and receives \$50 each month.

A tank begins full and drains at a constant rate.

A flat monthly membership fee does not change with the number of visits.

The first graph rises and begins above zero.

The second graph falls.

The third graph is horizontal.

Before writing an equation, describe the movement.

Starts at:

Changes by:

Increases or decreases:

Input:

Output:

For the savings account:

$$S = 50m + 200$$

The slope is 50 dollars per month. The intercept is \$200.

Distinguish Correlation From Cause

A graph may show that two quantities change together. That alone does not prove that one causes the other.

Suppose students who spend more time in a tutoring center also have higher course scores. The relationship may reflect tutoring, motivation, prior preparation, course difficulty, or other factors.

A placement question may ask only for the trend. Answer the mathematical question without claiming more than the data show.

Use precise language:

"As study time increases, the score tends to increase in this data."

Avoid:

"Study time is the only reason the score increased."

Read Piecewise Situations

Some rules change after a threshold.

A parking garage charges \$8 for the first two hours and \$3 for each additional hour.

The cost does not follow one simple rate from zero.

For three hours:

$$8 + 3(1) = 11$$

For five hours:

$$8 + 3(3) = 17$$

A graph may show a flat or different section before the rate changes.

Read the domain of each rule.

Data Choice Challenge

Create a small data set with one outlier:

$$6, 7, 7, 8, 9, 30$$

Calculate mean and median.

Mean:

$$\frac{67}{6} \approx 11.17$$

Median:

$$\frac{7+8}{2} = 7.5$$

Write one sentence explaining which better represents a typical value and why.

Then remove the outlier and recalculate. Notice how the mean changes more strongly.

Use Intercepts as Events

An x intercept occurs where $y = 0$.

A y intercept occurs where $x = 0$.

In a cost context, the y intercept may be a starting fee.

In a distance context, an x intercept may represent the time when a remaining distance reaches zero.

Do not describe an intercept only as a point. Translate what the zero means.

Find a Missing Table Value

Suppose the relationship is linear.

x	y
1	7

x	y
2	11
3	15
5	?

The output rises by 4 when the input rises by 1.

From $x = 3$ to $x = 5$ is an increase of 2, so y rises by 8.

$$15 + 8 = 23$$

The missing value is 23.

Formula:

$$y = 4x + 3$$

Check:

$$4(5) + 3 = 23$$

Interpret Domain

The domain is the set of allowed input values.

A formula may permit all real numbers mathematically while the situation permits only some.

If x represents the number of courses, negative values do not fit. If only whole courses can be taken, noninteger values may not fit either.

Context can restrict the domain beyond the algebraic rule.

In a Nutshell

Graphs, tables, formulas, and descriptions can represent the same relationship.

Read the scale before the shape. Identify input and output. Interpret slope as a rate with units. Recognize the starting value. Use function notation correctly. Choose mean or median with attention to the data.

> *Do not ask only what the graph looks like. Ask what each unit of movement means.*

Chapter 12: Solve Word Problems Without Guessing

Tasha read the problem twice and underlined every number.

A program charged a registration fee of \$45 plus \$18 per class. The total bill was \$153. How many classes were included?

Tasha multiplied all three numbers because multiplication often appeared in cost problems. The result was enormous.

On the second attempt she asked a better question:

What is unknown?

The number of classes.

What relationship is described?

Fixed fee plus cost per class equals total cost.

She wrote:

$$45 + 18c = 153$$

Then:

$$18c = 108$$
$$c = 6$$

The words had not hidden an arithmetic trick. They had described an equation.

The chapter principle is:

> ***Do not search a word problem for a magic keyword. Build a representation of the situation, then choose the mathematics that matches it.***

Use Read Represent Solve Check

This book uses four steps.

Read

Identify what is known and what is being asked.

Represent

Use a diagram, table, equation, ratio, or labeled list.

Solve

Carry out the mathematics one step at a time.

Check

Test the answer against the original situation, including units and reasonableness.

This method is slower than guessing for the first few problems. It becomes faster because it prevents restarts.

Identify the Unknown

Write a sentence before an equation:

Let x represent...

Example:

A total of 72 credits must be completed. A student has finished 39. How many remain?

Let x represent remaining credits.

$$39 + x = 72$$
$$x = 33$$

Answer: 33 credits remain.

The unit belongs in the answer.

Separate Relevant From Irrelevant Information

A problem may include information that does not affect the solution.

Example:

A 38-year-old student drives 24 miles each way to campus three days per week. Her car averages 32 miles per gallon. How many miles does she drive to campus and back each week?

Age and fuel efficiency are irrelevant to the requested distance.

Round trip distance:

$$24 \times 2 = 48$$

Weekly distance:

$$48 \times 3 = 144$$

Answer: 144 miles.

Do not force every number into the calculation.

Heather's prevention rule was:

> ***Every number must earn its place by helping answer the final sentence.***

Treat Units as Clues

Suppose a worker earns \$19.50 per hour for 7.5 hours.

Desired unit: dollars.

$$\frac{\$19.50}{hour} \times 7.5\ hours$$

Hours cancel, leaving dollars.

Suppose a trip covers 180 miles at 60 miles per hour.

Desired unit: hours.

$$180\ miles \div 60\frac{miles}{hour} = 3\ hours$$

If the units do not lead to the requested unit, reconsider the setup.

Avoid Keyword Traps

Keywords can help, but they are not reliable by themselves.

"More than" may indicate addition:

A fee is \$20 more than another fee.

"More than" may also appear in a comparison that requires subtraction:

How much more is \$75 than \$48?

$$75 - 48 = 27$$

"Of" often means multiplication, especially in percent problems:

20 percent of 60:

$$0.20 \times 60 = 12$$

But reading only one word can still miss the larger structure.

Ask what relationship is happening.

Solve Consecutive Number Problems

Three consecutive integers have a sum of 72. Find them.

Let the first integer be x.

Then the next two are:

$$x + 1$$
$$x + 2$$

Equation:

$$x + (x + 1) + (x + 2) = 72$$
$$3x + 3 = 72$$
$$3x = 69$$
$$x = 23$$

The integers are 23, 24, and 25.

Check:

$$23 + 24 + 25 = 72$$

Solve Distance Rate Time Problems

Use:

$$d = rt$$

A vehicle travels at 55 miles per hour for 3.5 hours. Find distance.

$$d = 55(3.5) = 192.5$$

Answer: 192.5 miles.

A vehicle travels 270 miles in 4.5 hours. Find average rate.

$$r = \frac{d}{t}$$

$$r = \frac{270}{4.5} = 60$$

Answer: 60 miles per hour.

Keep units attached.

Solve Work and Production Problems

A team processes 84 forms in six hours at a constant rate. How many forms can it process in nine hours?

Unit rate:

$$84 \div 6 = 14$$

Then:

$$14 \times 9 = 126$$

Answer: 126 forms.

A different problem may include a fixed starting amount or changing rate. Do not assume every work situation is proportional.

Solve Mixture Style Problems

Basic placement questions may ask about combining quantities with different values.

Example:

A student has quiz scores of 72, 81, 85, and 90. What score is needed on a fifth quiz for an average of 84?

Five quizzes with an average of 84 require a total of:

$$5 \times 84 = 420$$

Current total:

$$72 + 81 + 85 + 90 = 328$$

Needed score:

$$420 - 328 = 92$$

Answer: 92.

Check:

$$\frac{328+92}{5} = 84$$

Solve Percent Applications

A course fee of $600 increases by 7 percent.

Increase:

$$0.07 \times 600 = 42$$

New fee:

$$600 + 42 = 642$$

A $600 fee is reduced by 7 percent.

Decrease:

$$42$$

New fee:

$$600 - 42 = 558$$

Circle whether the question asks for:

the amount of change
the original amount
the new amount
the percent of change
These are different unknowns.

Solve Geometry Applications

Rectangle perimeter:

$$P = 2l + 2w$$

Rectangle area:

$$A = lw$$

A rectangle has length 12 feet and width 7 feet.

Perimeter:

$$P = 2(12) + 2(7) = 38\ feet$$

Area:

$$A = 12(7) = 84\ square\ feet$$

Do not mix linear units and square units.

A perimeter measures distance around.
An area measures surface covered.

Build a Table for Multistep Problems

A table can prevent confusion when several rates or choices are involved.

Example:
Plan A charges \$25 plus \$8 per session.
Plan B charges \$10 plus \$11 per session.
When are the costs equal?

Plan	Fixed fee	Per session	Total
A	25	8	$25 + 8s$
B	10	11	$10 + 11s$

Set equal:

$$25 + 8s = 10 + 11s$$
$$15 = 3s$$
$$s = 5$$

Check:

Plan A:

$$25 + 8(5) = 65$$

Plan B:

$$10 + 11(5) = 65$$

The plans cost the same at five sessions.

Check the Context

An algebraic answer can be mathematically correct and contextually impossible.

If you solve for the number of classes and obtain 6.4, ask whether partial classes are allowed.

If a length is negative, the setup or interpretation may be wrong.
If a person's age is 220, recheck.
If a probability is greater than 1 or less than 0, recheck.
If a discount produces a price higher than the original, recheck.
Context is part of the answer.

Word Problem Practice

1. A student has completed 18 of 60 required credits. How many remain?

2. A subscription costs $12 per month plus a one-time $25 setup fee. Find the first year cost.
3. A car travels 315 miles in 5.25 hours. Find the average speed.
4. A $240 course fee is discounted by 15 percent. Find the sale price.
5. Three consecutive integers have a sum of 96. Find them.
6. A rectangle has perimeter 54 inches and length 16 inches. Find the width.
7. A worker earns $21 per hour and receives a $75 weekly bonus. How many hours produce total weekly pay of $810?
8. Four test scores are 78, 82, 89, and 91. What fifth score gives an average of 86?
9. Plan A costs $50 plus $6 per use. Plan B costs $20 plus $9 per use. When are they equal?
10. A recipe uses 2.5 cups for eight servings. How many cups are needed for 20 servings?

Answers:

1. 42 credits
2. $169
3. 60 miles per hour
4. $204
5. 31, 32, and 33
6. 11 inches
7. 35 hours
8. 90
9. 10 uses
10. 6.25 cups

For question 6:

$$54 = 2(16) + 2w$$
$$54 = 32 + 2w$$
$$22 = 2w$$
$$w = 11$$

For question 7:

$$21h + 75 = 810$$
$$21h = 735$$
$$h = 35$$

Your Representation Challenge

Choose three missed word problems from the diagnostic or practice sets.

For each one, create two representations before solving.

Possible pairs:

equation and diagram

table and equation

ratio and unit analysis

formula and labeled sketch

Then compare which representation made the relationship clearest.

The goal is not to draw for decoration. The goal is to externalize structure.

Use a Representation Decision

When the problem describes equal rates, try a unit rate or proportion.

When it describes a fixed amount plus a changing amount, try a linear expression.

When it asks about a shape, draw and label it.

When it gives several categories or time periods, try a table.

When it asks for a missing quantity in a known relationship, write an equation or formula.

This is not a rigid decision tree. It is a starting menu.

Luis once used a proportion for a service with a fixed fee. The answer failed the context check. He rewrote the situation as:

$$C = 12h + 30$$

The representation changed because the relationship was not proportional.

Rewrite the Final Question

Before solving, restate the request in your own words.

Original:

“What is the total amount paid after applying a 15 percent discount to the original $240 charge?”

Rewrite:

“I need the final price, not the discount amount.”

Now the work has two stages.

Discount:

$$0.15(240) = 36$$

Final price:

$$240 - 36 = 204$$

This short rewrite prevents an answer that stops too soon.

Solve From the End

Some problems are easier when you work backward.

After a \$25 fee is added, the total is \$145. What was the amount before the fee?

$$x + 25 = 145$$

$$x = 120$$

After a 20 percent increase, a quantity is 96.

$$1.20x = 96$$

$$x = 80$$

Working backward does not mean reversing every number blindly. It means identifying the operation that produced the final state and applying its inverse.

Write an Answer Sentence

Finish with a complete sentence containing the unit and context.

Weak:

6

Strong:

The bill includes six sessions.

The sentence forces one last check. If the number does not fit the noun, the setup may be incomplete.

Check With a Second Method

When time allows, solve a word problem two ways.

A \$90 item is discounted by 20 percent.

Method one:

$$0.20(90) = 18$$

$$90 - 18 = 72$$

Method two:

$$90(0.80) = 72$$

Agreement increases confidence.

For a rate problem, use both unit rate and proportion.

Different methods are especially useful when the result affects several later answers.

Label Diagrams

A diagram without labels can create new confusion.

For a rectangle, mark length and width.

For a distance problem, mark rate, time, and direction.

For a percent bar, mark the whole as 100 percent and the known part.

A rough diagram is enough. Accuracy of relationship matters more than artistic scale.

Reject Impossible Answers

Create a short impossible answer list:

A probability below zero or above one.

A negative physical length.

A percentage discount that makes the price rise.

A number of people expressed as a noninteger when partial people are impossible.

A time longer than the total period described.

The context check is part of solving, not an optional final decoration.

In a Nutshell

Word problems are translation problems.

Read for the unknown. Separate relevant information. Represent the relationship. Solve one step at a time. Check units, size, and context.

Do not trust a keyword more than the whole situation.

> *The sentence tells a story. Your representation tells you which mathematics the story requires.*

Chapter 13: Add Higher Math Only When Needed

Advanced topics earn a place in this plan only when the official test outline or Target Course requires them. With a short deadline, studying every impressive-looking topic can displace the arithmetic and algebra that matter more.

> ***Higher mathematics belongs in your plan only when the official test outline or Target Course requires it. Advanced does not automatically mean high value.***

Use Exponent Rules

An exponent shows repeated multiplication.

$$x^4 = x \times x \times x \times x$$

Product rule:

$$x^a x^b = x^{a+b}$$

Example:

$$x^3 x^5 = x^8$$

Quotient rule:

$$\frac{x^a}{x^b} = x^{a-b}$$

when $x \neq 0$.

Example:

$$\frac{x^7}{x^2} = x^5$$

Power rule:

$$\left(x^a\right)^b = x^{ab}$$

Example:

$$\left(x^3\right)^4 = x^{12}$$

Power of a product:

$$(ab)^n = a^n b^n$$

Zero exponent:

$$x^0 = 1$$

for $x \neq 0$.

Negative exponent:

$$x^{-n} = \frac{1}{x^n}$$

Example:

$$2^{-3} = \frac{1}{2^3} = \frac{1}{8}$$

Do not add exponents when adding terms.

$$x^2 + x^3$$

does not equal x^5. Exponent rules apply to multiplication and division structures, not every nearby pair of powers.

Work With Scientific Notation

Scientific notation has the form:

$$a \times 10^n$$

where $1 \leq |a| < 10$.

Write 4,800,000 in scientific notation:

$$4.8 \times 10^6$$

Write 0.00072:

$$7.2 \times 10^{-4}$$

Multiply:

$$(3 \times 10^4)(2 \times 10^3)$$
$$= 6 \times 10^7$$

Divide:

$$\frac{8 \times 10^6}{2 \times 10^2}$$
$$= 4 \times 10^4$$

Check that the leading factor remains between 1 and 10.

Understand Radicals

A square root asks which nonnegative number squares to produce the value.

$$\sqrt{49} = 7$$

because:

$$7^2 = 49$$

Simplify:

$$\sqrt{72}$$

Factor out the largest perfect square:

$$\sqrt{(36 \times 2)}$$
$$= 6\sqrt{2}$$

Add like radicals:

$$3\sqrt{5} + 2\sqrt{5} = 5\sqrt{5}$$

You cannot combine:

$$\sqrt{2} + \sqrt{3}$$

because the radical parts differ.

Remember that:

$$\sqrt{(x^2)} = |x|$$

for real x. The principal square root is nonnegative.

Add and Multiply Polynomials

A polynomial is an expression containing terms with nonnegative integer exponents.

Add:

$$\left(3x^2 + 2x - 5\right) + \left(x^2 - 7x + 4\right)$$

Combine like terms:

$$4x^2 - 5x - 1$$

Multiply a monomial:

$$3x\left(2x^2 - 5x + 4\right)$$
$$= 6x^3 - 15x^2 + 12x$$

Multiply two binomials:

$$(x + 3)(x + 5)$$

Distribute every term:

$$x^2 + 5x + 3x + 15$$
$$= x^2 + 8x + 15$$

A box or area model can help keep every product visible.

Factor Common Factors First

Factoring reverses multiplication.

Factor:

$$6x^2 + 9x$$

The greatest common factor is $3x$:

$$3x(2x + 3)$$

Check by distributing.

Always look for a common factor before trying other methods.

Factor Trinomials

Factor:

$$x^2 + 8x + 15$$

Find two numbers that multiply to 15 and add to 8.

The numbers are 3 and 5.

$$x^2 + 8x + 15 = (x + 3)(x + 5)$$

Factor:

$$x^2 - x - 12$$

Find two numbers that multiply to -12 and add to -1.

The numbers are -4 and 3.

$$(x - 4)(x + 3)$$

Check by multiplying.

Solve Quadratic Equations

A quadratic equation can often be solved by factoring.

$$x^2 + 8x + 15 = 0$$

Factor:

$$(x + 3)(x + 5) = 0$$

Use the zero product property:

$$x + 3 = 0$$

or:

$$x + 5 = 0$$

So:

$$x = -3, \quad x = -5$$

A product equals zero when at least one factor equals zero.

For a quadratic equation $ax^2 + bx + c = 0$, where $a \neq 0$, the quadratic formula may apply when factoring is difficult or unavailable:

$$x = (-b \pm \sqrt{(b^2 - 4ac)}) / 2a$$

Example:

$$x^2 - 2x - 1 = 0$$

Here $a = 1, b = -2, c = -1$.

$$x = [-(-2) \pm \sqrt{((-2)^2 - 4(1)(-1))}] / [2(1)]$$

$$x = (2 \pm \sqrt{8}) / 2$$

$$x = 1 \pm \sqrt{2}$$

Use parentheses carefully when substituting negative coefficients.

Understand the Discriminant

The expression:

$$b^2 - 4ac$$

is the discriminant.

For a quadratic with real coefficients, a positive discriminant gives two distinct real solutions. A zero discriminant gives one repeated real solution. A negative discriminant gives no real solutions, although it gives two nonreal complex solutions.

This may appear on advanced placement tests. Use it only if your official outline includes quadratics.

Review Function Features

For:

$$f(x) = x^2 - 4$$

Find $f(3)$:

$$f(3) = 9 - 4 = 5$$

Solve $f(x) = 0$:

$$x^2 - 4 = 0$$
$$(x - 2)(x + 2) = 0$$
$$x = 2, \quad x = -2$$

These values are the x intercepts of the graph.

For a linear function, rate of change is constant. For a quadratic function, the graph is a parabola and the rate of change is not constant.

Use Geometry Formulas Carefully

Common formulas may include:

Rectangle:

$$A = lw$$

Triangle:

$$A = \frac{1}{2}bh$$

Circle:

$$C = 2\pi r$$
$$A = \pi r^2$$

Right triangle:

$$a^2 + b^2 = c^2$$

where c is the hypotenuse.

Example:

A right triangle has legs 6 and 8.

$$6^2 + 8^2 = c^2$$
$$36 + 64 = c^2$$
$$100 = c^2$$
$$c = 10$$

Use positive length.

Know What to Skip

Stop before adding a topic when you cannot answer yes to at least one of these:

1. The official test outline names it.
2. The official sample questions use it.
3. The Target Course lists it as a prerequisite.
4. Your diagnostic shows it is a Placement Gap.
5. An adviser or department has confirmed it matters.

This protects Foundation Skills from being crowded out by advanced review.

Higher Math Practice

1. Simplify $x^4 x^7$.
2. Simplify $\frac{a^9}{a^3}$.
3. Simplify $\left(2x^3\right)^2$.
4. Write 0.000056 in scientific notation.
5. Simplify $\sqrt{108}$.
6. Add $\left(2x^2 + 5x - 1\right) + \left(3x^2 - 2x + 7\right)$.
7. Multiply $(x + 4)(x - 3)$.
8. Factor $8x^2 - 12x$.
9. Factor $x^2 + 9x + 20$.
10. Solve $x^2 - 5x + 6 = 0$.

11. Find $f(-2)$ for $f(x) = 2x^2 - 3x + 1$.
12. A right triangle has legs 9 and 12. Find the hypotenuse.

Answers:

1. x^{11}
2. a^6
3. $4x^6$
4. 5.6×10^{-5}
5. $6\sqrt{3}$
6. $5x^2 + 3x + 6$
7. $x^2 + x - 12$
8. $4x(2x - 3)$
9. $(x + 4)(x + 5)$
10. $x = 2$ or $x = 3$
11. 15
12. 15

For question 11:

$$f(-2) = 2(-2)^2 - 3(-2) + 1$$
$$= 2(4) + 6 + 1 = 15$$

Use the Advanced Topic Gate

Before studying a Higher-Math topic, answer four questions.

Is it named in the official content outline?

Does it appear in official sample questions?

Does the Target Course use it early?

Are the prerequisite skills stable?

If the first three answers are no, postpone the topic.

If the fourth answer is no, repair the prerequisite first.

For example, factoring a quadratic depends on integer factors, signs, multiplication, and combining like terms. A learner who is still rebuilding signed multiplication should not interpret factoring difficulty as one giant advanced gap. The First Broken Link may be much earlier.

Connect Factoring and Graphs

Consider:

$$y = x^2 - 5x + 6$$

Factor:

$$y = (x - 2)(x - 3)$$

The zeros occur when:

$$x = 2$$

or:

$$x = 3$$

These are the x intercepts of the graph.

The algebraic form and graph tell the same story in different ways.

This connection matters more than memorizing a long list of isolated factoring exercises.

Check Exponent Expressions by Expansion

Simplify:

$$\left(x^2\right)^3$$

The power rule gives:

$$x^6$$

Check by expansion:

$$\left(x^2\right)^3 = x^2 x^2 x^2 = x^6$$

Now compare:

$$x^2 + x^2 + x^2 = 3x^2$$

Addition counts like terms. Multiplication combines exponents.

This distinction prevents one of the most common advanced algebra errors.

Choose Exact or Approximate Form

$$\sqrt{2}$$

is an exact value.

A decimal such as 1.414 is an approximation.

If a question asks for exact form, keep the radical.

If it asks for a rounded decimal, calculate and round as directed.

Likewise:

$$\frac{1}{3}$$

is exact, while 0.333 is approximate.

Read the requested answer form before changing representations.

Higher Math Repair Plan

Choose one required advanced topic.

Write:

Prerequisite skills:

First Broken Link:

One worked example:

Two focused questions:

One Mixed Practice question:

Delayed retest date:

This keeps Higher-Math study connected to the same Five-Rung Placement Ladder used throughout the book.

Rational Expressions Need Restrictions

Consider:

$$\frac{x+3}{x-2}$$

The denominator cannot equal zero.

$$x - 2 \neq 0$$

$$x \neq 2$$

Any simplification or equation solving must preserve that restriction.

Example:

$$\frac{x^2-4}{x-2}$$

Factor:

$$\frac{(x-2)(x+2)}{x-2}$$

This simplifies to:

$$x + 2$$

but only when $x \neq 2$. The original expression was undefined at 2.

Use this topic only if your verified route includes rational expressions.

Complete the Square Concept

Consider:

$$x^2 + 6x$$

Half of 6 is 3. Square it to get 9.

$$x^2 + 6x + 9 = (x + 3)^2$$

This process creates a perfect square trinomial.

To solve:

$$x^2 + 6x = 7$$

Add 9 to both sides:

$$x^2 + 6x + 9 = 16$$
$$(x + 3)^2 = 16$$
$$x + 3 = \pm 4$$
$$x = 1 \; or \; x = -7$$

The method shows why both positive and negative square roots matter.

Read a Quadratic Graph

For:

$$y = (x - 2)^2 - 4$$

The vertex is $(2, -4)$.

The graph opens upward because the squared term has a positive coefficient.

Set $y = 0$ to find intercepts:

$$(x - 2)^2 - 4 = 0$$
$$(x - 2)^2 = 4$$
$$x - 2 = \pm 2$$
$$x = 0 \; or \; x = 4$$

Algebra and graph features reinforce each other.

Stop at the Verified Boundary

A broad placement platform may contain topics beyond the score band you need. Do not interpret every topic in the learning module as equally urgent.

Return to the Target Course, official score guidance, and diagnostic.

The Higher-Math Route is still targeted preparation, not a complete precalculus textbook.

Preserve Core Review

For every two Higher-Math sessions, include one short Mixed Practice set from the core chapters. Advanced study should not allow fractions, signs, equations, or word problem translation to fade.

The placement system may test a broad range. Strong advanced work still depends on reliable foundations.

In a Nutshell

Advanced mathematics is useful when it serves your Target Course.

Use exponent rules according to the operation. Read and write scientific notation. Simplify radicals by factoring out perfect squares. Combine like polynomial terms. Distribute to multiply. Factor as the reverse of multiplication. Solve quadratics by factoring or another method when required.

Do not study an advanced topic to prove that you are serious.

> ***The serious plan protects the skills that matter most.***

Chapter 14: Practice Like the Test

Topic labels make method selection easy: a page marked "percentages" announces the tool before the problem begins. A placement test removes that cue. Mixed Practice is where you learn to identify the method as well as execute it.

> *Topic practice teaches you how to use a method. Mixed Practice teaches you when to use it. Placement readiness requires both.*

Move From Blocked to Mixed Practice

Blocked practice groups similar problems.

Examples:

Ten fraction additions Ten percentage conversions Ten one step equations

This is useful when learning a new procedure. You can focus on the steps without repeatedly deciding which method applies.

Mixed Practice combines topics.

Examples:

A fraction problem A graph question A percentage decrease An equation A rate problem

Now you must identify the structure before solving.

Use both forms.

A good progression is:

1. Worked examples
2. Similar focused problems
3. A changed version
4. Mixed problems
5. A delayed retest

Do not remain in blocked practice because it feels smoother. Smooth practice can create confidence that disappears when labels are removed.

Match the Question Format

If your test uses multiple choice, practice producing an answer before looking at the choices when possible.

Then use the options strategically:

eliminate wrong sign

eliminate impossible size

check units
test a simple value
substitute choices when efficient
Do not let answer choices replace learning. Recognition can feel like recall.
If your test uses open response, practice entering answers exactly.
Pay attention to:
negative signs
fraction format
decimal rounding
percent form
exponents
ordered pairs
interval or inequality notation if required
ALEKS PPL uses an adaptive, open-response format, so students should not prepare only through multiple-choice recognition (McGraw Hill, n.d.-b).

Understand Adaptive Difficulty

An adaptive test uses previous responses to help select later questions. Harder questions do not automatically mean you are failing. Easier questions do not automatically reveal your final result.

McGraw Hill states that the current ALEKS PPL placement assessment evaluates knowledge across 314 topics in 25 questions or fewer (McGraw Hill, n.d.-a). A separate current official ALEKS PPL page describes the assessment as adaptive and open response (McGraw Hill, n.d.-b). Florida describes PERT as computer adaptive (Florida Department of Education, n.d.).

Your task is not to reverse engineer the algorithm during the test.

Use the one question rule:

> *This question is the only question whose answer I can change right now.*

Do not estimate your score from the last item.

Practice Calculator Rules

If your verified test rules limit calculator use, your practice must reflect that.

College Board states that standard ACCUPLACER mathematics testing does not permit a personal handheld calculator. Some questions display an onscreen calculator, with accommodations handled separately (College Board, n.d.-a).

For ACCUPLACER practice:

complete basic arithmetic without a handheld calculator

use estimation

practice the official calculator tutorial

do not assume every problem offers a calculator

For another assessment, follow its current official rule.

Even when a calculator is allowed, write the setup first. A calculator can evaluate the wrong expression perfectly.

Carlos entered:

$$45 + 18 \times 6$$

and obtained 153.

That was useful because the equation had already been built. If he had entered all visible numbers without a relationship, the calculator would not have corrected the thinking.

Organize Scratch Work

Divide scratch paper into numbered boxes.

Write the question number.

Copy only essential information.

Keep one transformation per line.

Circle the final answer.

For multiple choice, write the selected letter and the computed value.

For word problems, write units.

For skipped items on a test that permits review, place a clear mark beside the number.

Messy work is not a personality trait. It is a preventable source of error.

Renee often solved one problem across three corners of the page. Her numbers from different questions began to mix. A simple box around each problem reduced copying errors immediately.

Use a Three Pass Review

After a practice assessment, do not look only at the score.

Use three passes.

Pass One Correctness

Mark each answer correct or incorrect.

Pass Two Cause

Classify each error:

unknown concept
forgotten rule
wrong method selected
arithmetic
sign
copying
units
misread question
time or fatigue
answer entry

Pass Three Repair

Choose the next action:

relearn the concept
solve two similar problems
write a prevention rule
practice the interface
ask for help
add the skill to delayed review

A score tells you how the attempt ended. The repair pass tells you what the attempt is worth.

Practice Under Realistic Conditions

Your first attempt at a skill should not happen under test conditions. Your later rehearsal should.

For Practice Assessment One:

use a quiet space
follow verified calculator rules
avoid notes
use the official question format when possible
record time without forcing speed
complete the Error Pattern Log afterward

For Practice Assessment Two:

use the expected time or pacing conditions

prepare identification and permitted materials as a rehearsal
use the Calm Test Routine
avoid pausing for instruction
review only after the assessment is complete

College Board states that, in most instances, ACCUPLACER is untimed, but the school or testing center administering the test will confirm local conditions (College Board, n.d.-f). Your appointment window and concentration still create practical limits.

Build Pacing Without Rushing

Use a pace check, not a panic clock.

After a group of five questions, ask:
Am I reading every final sentence?
Am I writing enough work to check?
Have I spent too long on one item?
Is fatigue changing my accuracy?

Your assessment may or may not permit backtracking. Verify the official tutorial or the institution's written instructions before test day, then practice with the same rule.

For an untimed adaptive assessment, do not interpret "untimed" as permission to stay trapped on one problem indefinitely. Use a reasonable attempt, check, and submit when you have done the work available to you.

Learn From Wrong Answers

When reviewing an incorrect multiple choice problem, do not stop after reading the correct solution.

Ask why each wrong option might attract someone.
Example:
What is 20 percent of 80?
A. 4 B. 16 C. 60 D. 100
Correct: B.
Option A may come from dividing 80 by 20.
Option C may come from subtracting 20.
Option D may come from adding 20.
This analysis exposes Error Patterns.

Some multiple-choice questions use plausible distractors based on common errors. Reviewing why each option is wrong can expose an Error Pattern without assuming that every assessment uses the same design.

Know When a Topic Is Ready

Stop focused review when you can:

1. Solve the skill accurately in blocked practice.
2. Explain the method.
3. Solve a changed version.
4. Recognize it inside Mixed Practice.
5. Repeat the success after a delay.

If the skill fails only under fatigue, the repair may involve pacing and breaks rather than more explanation.

If the skill fails when the wording changes, the repair may involve translation.

If the skill fails after two days, the repair may involve spaced retrieval.

Ready does not mean you will never miss another question. It means the skill is reliable enough to stop consuming most of your time.

Practice Assessment One Protocol

Use the assessment in Appendix D.

Before starting, write:

> ***My goal is to collect accurate evidence, not protect my confidence.***

After finishing:

score by domain

classify every error

identify the three highest value repairs

schedule those repairs within 48 hours

retest the same skill with new problems

do not retake the entire assessment immediately

An immediate retake can measure memory of the test rather than learning.

Practice Assessment Two Protocol

Use Appendix E near the end of the plan.

Before starting, verify your test date, format, calculator rule, and Target Course.

After scoring, ask:

Did my highest value errors decrease?
Which errors are content gaps?
Which errors are process gaps?
Does my current performance support the Target Course?
What should I review lightly?
What should I stop trying to learn before test day?
Practice Assessment Two is a readiness rehearsal, not a final judgement.

Review a Miss From Start to Finish

Suppose you miss this problem:

A \$250 fee increases by 12 percent. Find the new fee.
Your answer is \$30.
Do not record only "wrong."
The \$30 is the amount of increase:

$$0.12(250) = 30$$

The question asks for the new fee:

$$250 + 30 = 280$$

Error Pattern: stopped before answering the requested quantity.

Prevention rule: circle "increase amount" or "new total."
Fresh retest:
A \$180 fee decreases by 15 percent. Find the new fee.
Decrease:

$$0.15(180) = 27$$

New fee:

$$180 - 27 = 153$$

The retest changes the numbers and direction. It checks the rule, not memory of the first answer.

Use Confidence Ratings Carefully

After each practice item, rate confidence from one to three.

One: mostly guessed or blank.
Two: method seemed right but needed checking.
Three: confident and able to explain.
Then compare confidence with correctness.

High confidence plus wrong answer deserves attention. It may reveal a stable misconception.

Low confidence plus correct answer may indicate a Recall Skill that needs another delayed attempt.

The goal is not to feel confident about everything. It is to make confidence better match actual performance.

Rehearse Recovery

During practice, intentionally use the Calm Test Routine after a difficult item.

Do not wait until test day to learn how to reset.

Finish the hard item. Put the pencil down. Exhale. Read the next question without carrying the previous one forward.

Recovery is a test skill.

> ***One difficult problem should cost one problem, not the next five.***

Create an Assessment Repair Sheet

After each practice assessment, create one page with three columns.

Keep

Skills that were accurate and need only maintenance.

Repair

High-Value Skills that caused errors.

Release

Topics outside the Target Course or already stable enough to stop receiving focused time.

This prevents the score report from turning into a broad panic review.

Practice Answer Entry

Write five answers in the form your test may request:

A negative integer:

$$-7$$

A fraction:

$$\frac{3}{8}$$

A decimal rounded to hundredths:

$$2.47$$

An ordered pair:

$$(4, -1)$$

An inequality:

$$x \leq 6$$

Check whether spaces, parentheses, fraction bars, or rounding directions matter in the official interface.

A correct calculation entered in the wrong form may still be marked wrong.

Stop Studying the Score

After reviewing an assessment, put the total aside and work from the repair list.

Repeatedly recalculating the percentage does not change readiness.

One corrected Error Pattern does.

Finish With One Transfer Item

After repairing a missed skill, solve one problem that looks different but uses the same relationship. Change the context, number form, or question wording.

Transfer is the evidence that you learned the method rather than memorized the correction.

In a Nutshell

Focused practice teaches methods. Mixed Practice teaches selection.

Match the test's question format, calculator rules, pacing, and interface. On adaptive tests, focus on the current item rather than interpreting difficulty. Organize scratch work. Review errors by cause and repair.

Do not waste a wrong answer by recording only that it was wrong.

> *The score measures one attempt. The Error Pattern Log turns the attempt into preparation.*

Chapter 15: Finish Calm and Decide Well

Aisha's test was scheduled for 9:00 on a Monday morning.

On Sunday afternoon she opened a new video about quadratic equations. Her Target Course did not require quadratics, and her diagnostic had not identified them as a gap. The video simply appeared beside something useful, and the title promised a shortcut.

Twenty minutes later Aisha was confused, tired, and suddenly certain that she knew nothing.

She closed the video and opened her one page Placement Brief.

Her test was ACCUPLACER Arithmetic and Quantitative Reasoning, Algebra, and Statistics. Her Target Course was a supported statistics section. Her remaining Error Patterns were sign mistakes and rushing percentage questions.

Quadratics were not the problem in front of her.

Aisha reviewed her two prevention rules, packed her identification, checked the testing location, and stopped studying.

The chapter principle is:

> ***The final week is not the time to prove how much more you can learn. It is the time to stabilize what matters, rehearse the test process, and prepare to make a sound decision after the result.***

Seven Days Before

Complete Practice Assessment Two.

Do not respond to the score with a broad return to every chapter. Identify the three highest value repairs.

A useful repair is specific.

Weak:

Review algebra.

Strong:

Practice equations with variables on both sides and check each answer by substitution.

Weak:

Work on fractions.

Strong:

Review addition with unlike denominators and solve four mixed questions after a one day delay.

Schedule the repairs across the next four days. Keep the final two days lighter.

Six Days Before

Review logistics.

Confirm:

date and start time

testing location or remote login

arrival time

identification

appointment confirmation

calculator rule

scratch paper rule

accessibility arrangements

transportation

parking or public transit

contact number for problems

College Board directs students to official sample questions, free practice tests, and test-taking tutorials before ACCUPLACER test day (College Board, n.d.-f).

Do not wait until the morning of the test to discover that your identification name does not match the registration or that remote testing requires a system check.

Five Days Before

Repair the first High-Value Skill.

Use this sequence:

One worked example Two focused problems One changed problem One Mixed Practice problem One delayed problem the next day

Stop if accuracy falls because of fatigue. A longer session is not valuable when it rehearses mistakes.

Four Days Before

Repair the second High-Value Skill.

Then review your Error Pattern Log from the beginning.

Count repeated patterns.

Example:

Sign error: seven early, one recent Misread final sentence: four early, three recent Decimal placement: three early, zero recent

This tells you where prevention rules are working.

Do not celebrate only correct topics. Celebrate reduced repeat errors.

Three Days Before

Repair the third High-Value Skill.

Complete a short mixed set of 10 questions. Do not take another full assessment unless your plan specifically requires it.

Review the official test interface.

If your test is adaptive, rehearse the one question rule.

If it is multiple choice, produce answers before checking options when possible.

If it is open response, practice exact entry.

Two Days Before

Stop adding new domains.

Review:

formula notes

conversion anchors

sign rules

Error Pattern prevention rules

Target Course

test details

Calm Test Routine

Complete three to five easy or moderate questions. The purpose is to keep movement familiar, not to create a final score.

Prepare clothing, transportation, food, medication taken as normally prescribed, and any permitted materials. This book cannot advise on personal medical needs. Follow your established care plan and testing accommodations.

One Day Before

Use light review only.

Do not retake every difficult question.

Do not compare scores with strangers online.

Do not change the Target Course because a commercial advertisement promises a higher placement overnight.

Write this statement:

> ***I will use the skills I have prepared. I do not need to know the answer to every question to produce useful evidence of my readiness.***

Then stop.

The Calm Test Routine

Use this routine before the first question and whenever you feel stuck.

Settle

Place both feet on the floor. Drop your shoulders. Exhale slowly.

Read

Read the full question without calculating.

Mark

Identify the requested quantity, units, signs, and restrictions.

Represent

Write an equation, diagram, table, estimate, or first step.

Solve

Work one line at a time.

Check

Ask whether the sign, size, unit, and context make sense.

Release

Submit or move on according to the test rules. Do not carry the previous question into the next one.

The routine does not guarantee calm. It gives you an action while anxiety is present.

When the First Question Feels Hard

A hard opening question can create a false story:

"I studied the wrong material."

"I am already failing."

"This is proof that I cannot do college math."

Replace the story with a task.

What is being asked?
What information is given?
What form should the answer take?
What is the smallest step I can write?

On an adaptive assessment, question difficulty can change in response to earlier answers. Do not infer a final score from the feeling of one item.

On a fixed assessment, one difficult item is still one item.

> *A hard question is a problem to work, not a forecast.*

Use Time Wisely

For an untimed test, take enough time to read and check. Do not confuse untimed with endless.

For a timed test, know the total time and question count. Build a rough pace, then preserve enough time for items you can solve.

If the system permits review, mark and return according to official instructions.

If it does not permit review, make the best supported decision and release the question.

Do not spend 15 minutes trying to protect one point while losing the chance to answer several later questions.

After the Test

Before interpreting the result, take a short break.

Then obtain the official score report or placement result. College Board provides a student process for retrieving ACCUPLACER scores and Skills Insight information for interpreting performance (College Board, n.d.-e; College Board, n.d.-j).

Record:

overall result
domain scores or diagnostic information
assigned course
Target Course comparison
retest eligibility
required learning modules
adviser instructions
deadlines
available support

Do not rely only on the number displayed immediately after testing. The course decision comes from the Placement Policy.

If You Reach the Target

Confirm enrollment steps.

Ask:

Is the placement valid for a limited time?

Must I register within a deadline?

Is a corequisite or lab attached?

What should I review before class?

What materials or software are required?

What tutoring is available from week one?

Then maintain skills.

Two or three short sessions each week can help keep relevant skills active between placement and the first day of class.

Use the Target Course readiness list, not random practice.

If You Do Not Reach the Target

Do not make the next decision while treating the score as a personal verdict.

Use the Decide rung.

Check:

1. Was the result used correctly under the current Placement Policy?
2. Are multiple measures available?
3. Is retesting allowed?
4. Is a learning module required?
5. Is corequisite support available?
6. Is the assigned course a prerequisite or a supported route?
7. How will the result affect the program timeline?
8. Which domain caused the Placement Gap?

A lower result can lead to several reasonable actions.

You may retest after targeted review.

You may request an adviser review.

You may document prior coursework.

You may choose a supported college level course.

You may accept a foundation course because the skill gap is real and important.

The book cannot choose among those options without your institutional details It can help you ask the right questions.

Retest With a New Plan

Do not retake immediately simply because the first score disappointed you.

First identify what will be different.

A useful retest plan includes:

a specific domain gap

a required learning module

a waiting period

new practice rather than memorized questions

a revised Error Pattern rule

realistic test conditions

a clear Target Course

McGraw Hill's standard ALEKS PPL description includes 12 months of access to a personalized learning module and up to five assessment attempts, but institutional settings can differ (McGraw Hill, n.d.-a).

Your college may set its own ACCUPLACER retest conditions. Ask the testing center for the current written rule.

Talk to an Adviser With Evidence

Bring:

Placement Brief

Target Course Map

score report

prior transcript

diagnostic summary

Error Pattern Log summary

program deadline

questions about support

Use a direct script:

"I am entering the ______ program. The first required mathematics course is ______. My placement result is ______. I would like to understand whether multiple measures, retesting, or a supported section are available, and which option gives me the strongest path through the program."

This is not asking for a favor. It is asking for a policy explanation.

Decide for Completion Not Pride

A course one level higher may look better. A course with the right support may produce better completion.

Ask:
Can I do the prerequisite skills independently?
How heavy is the rest of my semester?
Is the course compressed?
What support is built in?
Does this route delay a program application?
Will a lower course consume credit or aid without advancing the degree?
Will skipping foundation work create a larger risk later?
There is no universal answer.
Accurate placement remains the goal.

Final Readiness Check

Mark each statement yes, no, or uncertain.
I know the exact test or placement process.
I know the current calculator and retest rules.
I know my Target Course.
I can explain my three most common Error Patterns.
I have completed Mixed Practice.
I have rehearsed the test interface.
I know what identification and materials to bring.
I know how to retrieve the result.
I know who to contact about placement options.
I have a plan for maintaining skills before class.
Any "uncertain" answer becomes an action.

Create Two Post Test Plans

Before test day, write Plan A and Plan B.

Plan A: Target reached

Record how you will register, maintain skills, and use support in the first week.

Plan B: Target not reached

Record who you will contact, what policy you will verify, and how you will decide among retesting, multiple measures, corequisite support, or foundation work.

Preparing Plan B does not predict failure. It prevents a disappointed moment from becoming an unplanned decision.

Aisha's Plan B named the adviser, retest rule, and support section. She did not need every part of it, but knowing the options reduced the stakes she carried into the room.

Record the Test Experience

After testing, write brief notes before memory fades.

Which topics appeared?

Which format features surprised you?

Which Error Patterns appeared?

Which prevention rules worked?

How was your pacing?

Do not copy secure questions or violate test rules. Record your own process and broad skill areas.

These notes can guide a permitted retest or future course preparation.

End the 30 Days Deliberately

On the final evening, compare the first diagnostic page with your recent work.

Write three gains that are visible.

Write one gap that remains.

Write one support action.

Then close the placement plan.

Continuing to review indefinitely can become another form of avoidance. At some point, preparation must become performance and then decision.

In a Nutshell

The final week is for stabilization, not panic learning.

Complete the final assessment, repair three High-Value Skills, review the Error Pattern Log, confirm logistics, and rehearse the Calm Test Routine.

After the test, interpret the result through the Placement Policy and Target Course. Consider retesting, multiple measures, corequisite support, or foundation work with evidence rather than pride.

> *Test day ends with a score. Good placement planning continues with a decision.*

Conclusion: Your Placement Is a Starting Point

Maria returned to the kitchen table 31 days after she first closed the placement page

The laptop was in the same place. The yellow legal pad had changed.

The first pages contained hesitant calculations and questions written in large letters. The later pages showed equations with checks, fraction conversions, a short list of formulas, and an Error Pattern Log that had become less crowded.

Maria's test result placed her into the course she needed with an attached support lab.

A month earlier, she would have read the support lab as proof that she did not belong. Now she read the course description, compared it with her Target Course Map, and asked how the lab would fit her schedule.

The result was not perfect in the way panic had demanded. It was useful.

The final principle of this book is:

> ***A placement result is a starting point inside a larger plan. Your lasting achievement is not one score. It is knowing how to find a gap, rebuild a skill, use support, and keep moving.***

What You Have Built

You began by Clarifying.

You found the Placement Test, Placement Policy, Target Course, and local rules.

You Diagnosed.

You separated Ready Skills, Recall Skills, and Rebuild Skills.

You Rebuilt.

You worked through number sense, fractions, decimals, percentages, ratios, algebra, graphs, data, word problems, and any Higher-Math topics your route required.

You Rehearsed.

You used Mixed Practice, realistic test conditions, the Error Pattern Log, and the Calm Test Routine.

You Decided.

You interpreted the result through the course pathway instead of through shame.

That is the Five-Rung Placement Ladder completed once.

You can use it again.

Keep the Skills Active

A gap often exists between placement and the first day of class.

Use a maintenance plan.

Twice each week:

solve two Mixed Practice problems

review one earlier Error Pattern

complete one Target Course skill

explain one method without notes

Once each week:

complete a 15 minute mixed set

check the course syllabus or readiness materials

ask one unresolved question

Do not continue studying every topic from the placement assessment if the Target Course uses a narrower set. Shift from placement preparation to course preparation.

Use Support Early

Do not wait for the first failed exam to find the tutoring center.

Locate:

tutoring hours

mathematics lab

instructor office hours

peer study groups

online support

accessibility services

advising contacts

Visit once before you need urgent help.

Knowing the room, login, or appointment process reduces friction later.

A student who uses support is not admitting that placement was wrong. Support is part of how college works.

Keep the Error Pattern Log

The Error Pattern Log can continue into the course.

A new course may produce new patterns:

skipped units

incorrect calculator entry

weak graph interpretation
copied exponent
late start on assignments
misunderstanding of notation
Name the pattern. Write the prevention rule. Test the rule.
"Work harder" is still not a procedure.

Measure Progress With Evidence

Confidence may rise and fall.
Evidence is steadier.
Keep examples of what you can now do:
a diagnostic problem you once left blank
a fraction equation solved correctly
a graph interpreted with units
a word problem represented before calculation
an adviser question asked clearly
a test routine completed without rushing
When a later course feels difficult, do not use the feeling as proof that you have returned to the beginning.
Read the evidence.

Write the Continuation Plan

Complete this page.
My placement result:
My enrolled course:
My first class date:
The three skills I must maintain:
The two skills I still need to rebuild:
The support I will use in week one:
My weekly study blocks:
My most important Error Pattern prevention rule:
The date I will review this plan:
Then write one final sentence:

> ***My placement is the course where I begin, not the limit of where I can go.***

Principles to Remember

You did not need to relearn every mathematics topic you had ever met.

You needed to identify the right destination, rebuild the right skills, practice in the right format, and make the next decision with accurate information.

The Five-Rung Placement Ladder is not a promise that every reader will receive a particular score. It is a method for producing a better prepared attempt and a more informed choice.

You are allowed to be rusty.

You are also able to become ready.

> *The score opens one page. What you do next writes the chapter.*

Appendix A: No-Shame Diagnostic

Use this diagnostic before beginning the mathematics lessons.

Diagnostic limitation: These score bands and routing labels are practica study-planning heuristics created for this workbook. They have not beer standardized, normed, or validated as predictors of a college's placement decision Use them to organize review, not to estimate an official score or course placement.

Do not use notes, a search engine, or the answer section. Do not use a handheld calculator unless your verified Placement Test allows one and you have chosen tc reproduce that condition.

Write **C**, **U**, **P**, or **B** beside each response:

C: confident

U: unsure

P: needed a memory prompt

B: blank

Show your work.

Number Sense and Signed Numbers

1. Evaluate:

$$4,708 + 936$$

2. Evaluate:

$$6,000 - 2,759$$

3. Evaluate:

$$18 + 4(7 - 5)^2$$

4. Evaluate:

$$-8 + 13$$

5. Evaluate:

$$6 - 14$$

Fractions Decimals and Percentages

6. Simplify:

$$\frac{24}{36}$$

7. Add:

$$\frac{3}{5} + \frac{1}{4}$$

8. Multiply:

$$\frac{5}{8} \times \frac{12}{25}$$

9. Convert 0.075 to a percent.

10. Find 28 percent of 250.

Ratios Rates and Units

11. Simplify the ratio 18 to 30.

12. A worker completes 96 items in eight hours. Find the unit rate.

13. Solve:

$$\frac{5}{8} = \frac{x}{40}$$

14. Convert 4.5 feet to inches.

15. A map uses one inch to represent 25 miles. Two points are 3.6 inches apart. Find the real distance.

Algebraic Expressions and Equations

16. Simplify:

$$4x + 7x - 3$$

17. Distribute and simplify:

$$3(2x - 5) + x$$

18. Evaluate $2a^2 - 3a$ when $a =- 2$.

19. Solve:

$$5x + 7 = 32$$

20. Solve:

$$4x - 3 = 2x + 11$$

Graphs Data and Functions

21. Find the slope between $(2, 3)$ and $(6, 15)$.

22. Find $f(4)$ if:

$$f(x) = 3x - 5$$

23. Find the mean of:

$$6, 7, 7, 9, 11$$

24. Find the median of:

$$3, 5, 8, 12, 19, 24$$

25. A value rises from 120 to 150. Find the percent increase.

Word Problems and Higher Mathematics

26. A course costs $35 plus $18 per session. The total is $143. How many sessions were purchased?
27. A rectangle has length 11 inches and width 7 inches. Find its perimeter and area.
28. Three consecutive integers have a sum of 63. Find the integers.
29. Simplify:

$$x^3 x^5$$

30. Factor:

$$x^2 + 7x + 12$$

Diagnostic Answers

1. $5,644$
2. $3,241$
3. 34

$$7 - 5 = 2, \quad 2^2 = 4, \quad 4(4) = 16, \quad 18 + 16 = 34$$

4. 5
5. -8
6. $\frac{2}{3}$
7. $\frac{17}{20}$

$$\frac{3}{5} = \frac{12}{20}, \quad \frac{1}{4} = \frac{5}{20}$$

8. $\frac{3}{10}$

$$\frac{5}{8} \times \frac{12}{25} = \frac{1}{2} \times \frac{3}{5} = \frac{3}{10}$$

9. 7.5 percent
10. 70

$$0.28 \times 250 = 70$$

11. $3:5$
12. 12 items per hour
13. $x = 25$

14. 54 inches
15. 90 miles
16. $11x - 3$
17. $7x - 15$
18. 14

$$2(-2)^2 - 3(-2) = 8 + 6 = 14$$

19. $x = 5$
20. $x = 7$
21. 3

$$\frac{15-3}{6-2} = \frac{12}{4} = 3$$

22. 7
23. 8
24. 10

$$\frac{8+12}{2} = 10$$

25. 25 percent

$$\frac{150-120}{120} \times 100\% = 25\%$$

26. Six sessions

$$35 + 18s = 143$$
$$18s = 108$$
$$s = 6$$

27. Perimeter 36 inches, area 77 square inches
28. 20, 21, and 22
29. x^8
30. $(x + 3)(x + 4)$

Score by Domain

For each group of five:

Four or five accurate, explained answers: likely Ready or Recall.

Two or three accurate answers: mixed Recall and Rebuild.

Zero or one accurate answer: likely Rebuild.

Do not route yourself from this count alone. Review confidence marks and error causes.

A correct guessed answer does not establish a Ready Skill.

A wrong answer caused by one arithmetic slip does not automatically make the whole domain Rebuild.

Use Appendix B.

Appendix B: Diagnostic Routing Chart

The routes in this appendix are study-planning recommendations, not validated placement classifications. Combine them with the error analysis, the official test outline, and the Target Course. Your institution's Placement Policy remains controlling.

Foundation Route

Begin with the Foundation Route when any of these are true:

two or more errors in questions 1 through 10
difficulty with signed numbers
difficulty adding or multiplying fractions
difficulty converting decimals and percentages
inability to solve a one step equation
repeated basic arithmetic Error Patterns
Priority order:

1. Chapter 6
2. Chapter 7
3. Chapter 8
4. Chapter 9
5. Chapter 10
6. Chapter 12
7. Chapter 11
8. Chapter 14

Use Chapter 13 only if the Target Course requires it.

College Ready Route

Begin with the College-Ready Route when:

questions 1 through 15 are mostly accurate
arithmetic errors are occasional rather than structural
algebra, graphs, or word problems contain the main gaps
Mixed Practice is harder than labeled practice
Priority order:

1. Review missed parts of Chapters 6 through 8
2. Chapter 9

3. Chapter 10
4. Chapter 11
5. Chapter 12
6. Chapter 14
7. Chapter 15

Add Chapter 13 only when relevant.

Higher Math Route

Use the Higher-Math Route when:

core arithmetic and linear algebra are stable

the official test outline includes advanced algebra or functions

the Target Course is college algebra, precalculus, calculus, or another course requiring those skills

diagnostic questions 29 and 30 or related official questions reveal relevant gaps

Priority order:

1. Repair any core misses
2. Chapters 9 through 12
3. Chapter 13
4. Chapter 14
5. Chapter 15

Blended Route

Write each domain beside one label.

Number sense: Ready, Recall, or Rebuild

Fractions and percentages: Ready, Recall, or Rebuild

Ratios and units: Ready, Recall, or Rebuild

Algebra: Ready, Recall, or Rebuild

Graphs and data: Ready, Recall, or Rebuild

Word problems: Ready, Recall, or Rebuild

Higher mathematics: Ready, Recall, Rebuild, or Not Required

Spend about half your mathematics study time on Rebuild Skills, one third on Recall Skills, and the remaining time on Mixed Practice and maintenance. Adjust according to the Target Course.

Help Threshold

Use external help when:

a Rebuild Skill remains unclear after two focused sessions

the same Error Pattern appears three times after a prevention rule

you cannot identify the First Broken Link

the test date is within 10 days and the skill is central

anxiety prevents meaningful practice

Bring one worked example and one question you cannot solve.

Appendix C: Thirty-Day Study Calendars

Choose one time route. The topic order stays similar. The amount of practice changes.

Fifteen-Minute Route

Day 1: Verify Placement Test and Placement Policy.
Day 2: Build the Target Course Map.
Day 3: Take the No-Shame Diagnostic.
Day 4: Number sense and estimation.
Day 5: Signed numbers and order of operations.
Day 6: Fraction meaning and simplification.
Day 7: Fraction operations.
Day 8: Decimals and place value.
Day 9: Percent conversion and applications.
Day 10: Ratios and unit rates.
Day 11: Proportions and conversions.
Day 12: Algebra vocabulary and translation.
Day 13: Combining like terms.
Day 14: Distribution and substitution.
Day 15: One step equations.
Day 16: Multi step equations.
Day 17: Variables on both sides.
Day 18: Inequalities or formulas.
Day 19: Coordinate plane and slope.
Day 20: Tables and functions.
Day 21: Mean, median, and percent change.
Day 22: Word problem representation.
Day 23: Rate, percent, and geometry problems.
Day 24: Foundation repair or Higher-Math topic.
Day 25: Foundation repair or Higher-Math topic.
Day 26: Practice Assessment One, first half.
Day 27: Practice Assessment One, second half and review.
Day 28: Repair the three highest value errors.
Day 29: Practice Assessment Two selected questions and Calm Test Routine.
Day 30: Final Readiness Check and light review.

On assessment days, extend the session if the official assessment length requires it. A practice assessment cannot be meaningfully completed in 15 minutes. Split it across two sessions only for the first assessment. Complete the second under realistic conditions.

Thirty-Minute Route

Day 1: Verify test, calculator, timing, and retest rules.

Day 2: Build the Target Course Map and contact the college about missing facts.

Day 3: Take and score the No-Shame Diagnostic.

Day 4: Number sense, place value, estimation.

Day 5: Signed numbers, order of operations, factors.

Day 6: Fraction simplification, addition, subtraction.

Day 7: Fraction multiplication, division, mixed numbers.

Day 8: Decimals and conversion among forms.

Day 9: Percent part, whole, percent, increase, decrease.

Day 10: Ratios, unit rates, proportions.

Day 11: Unit conversion and scale.

Day 12: Algebra vocabulary and translation.

Day 13: Like terms and distribution.

Day 14: Substitution and formulas.

Day 15: One step and multi step equations.

Day 16: Variables on both sides and fraction equations.

Day 17: Inequalities and rearranging formulas.

Day 18: Coordinates, slope, linear equations.

Day 19: Tables, functions, input and output.

Day 20: Mean, median, range, and percent change.

Day 21: Read Represent Solve Check.

Day 22: Rates, costs, and percent word problems.

Day 23: Geometry, averages, and comparison problems.

Day 24: Route specific repair.

Day 25: Higher-Math topic or core Mixed Practice.

Day 26: Practice Assessment One.

Day 27: Error analysis and targeted repair.

Day 28: Practice Assessment Two.

Day 29: Final repair, interface, and logistics.

Day 30: Light review and Calm Test Routine.

Forty-Five-Minute Route

Use the 30 minute schedule and add:

five more independent problems on Learning Days

a 10 minute delayed review from an earlier topic

full official sample question sets where available

extra Higher-Math work on Days 23 through 25 when required

full assessment and review blocks on Days 26 and 28

Do not add random new topics. Extra time should deepen accuracy, transfer, and review.

Weekly Review

At the end of Days 7, 14, 21, and 28, answer:

I can now do:

My most common Error Pattern is:

The prevention rule that worked is:

The next High-Value Skill is:

A policy fact still unconfirmed is:

Missed Day Recovery

Miss one day: move the work to the next backup window.

Miss two days: complete a short review before new material.

Miss three or more days: retest three representative skills and revise the route.

Never restart from day one unless the test date has changed and a complete restart is genuinely useful.

Appendix D: Practice Assessment One

Complete this assessment after the core review chapters.

Use the calculator rule that matches your verified Placement Test. Show all scratch work. Questions 26 through 30 are Higher-Math questions. Complete them only if your test or Target Course requires them.

Questions

1. Evaluate:

$$7,204 - 3,867$$

2. Evaluate:

$$5 + 2(9 - 6)^2$$

3. Evaluate:

$$-12 + 7 - 4$$

4. Which is greater, 0.508 or 0.58?
5. Round 63.947 to the nearest hundredth.
6. Simplify:

$$\frac{42}{63}$$

7. Add:

$$\frac{5}{12} + \frac{7}{18}$$

8. Divide:

$$\frac{3}{5} \div \frac{9}{10}$$

9. Convert 0.032 to a percent.
10. A $480 fee is reduced by 17.5 percent. Find the new fee.
11. Simplify the ratio 28 to 49.
12. Seven identical items cost $94.50. Find the unit price.
13. Solve:

$$\frac{9}{14} = \frac{x}{56}$$

14. Convert 2.75 hours to minutes.
15. A drawing scale is one inch to 6.5 feet. A wall measures 4.2 inches on the drawing. Find the actual length.
16. Simplify:

$$8x - 3 + 5x + 11$$

17. Simplify:

$$-4(2y - 3) + 5y$$

18. Evaluate $3a^2 + 2a - 1$ when $a = -3$.
19. Solve:

$$6x - 5 = 31$$

20. Solve:

$$3x + 17 = 7x - 11$$

21. Solve:

$$\frac{x}{6} + \frac{1}{4} = \frac{11}{12}$$

22. Find the slope between $(-2, 5)$ and $(4, -7)$.
23. The data are 4, 6, 6, 9, 10, 13. Find the mean and median.
24. A service charges $22 plus $14 per visit. A bill is $120. How many visits were included?
25. A rectangle has area 96 square feet and width 8 feet. Find its length and perimeter.
26. Simplify:

$$\frac{x^8 x^3}{x^5}$$

27. Write 0.0000049 in scientific notation.
28. Simplify:

$$\sqrt{147}$$

29. Factor:

$$x^2 - 2x - 35$$

30. Solve:

$$x^2 + x - 12 = 0$$

Answers and Explanations

1. $3,337$

$$7,204 - 3,867 = 3,337$$

Estimate: $7,200 - 3,900 \approx 3,300$.

2. 23

$$9 - 6 = 3$$

$$3^2 = 9$$
$$5 + 2(9) = 23$$

3. -9

$$-12 + 7 = -5$$
$$-5 - 4 = -9$$

4. 0.58

Write 0.58 as 0.580. Then $0.580 > 0.508$.

5. 63.95

The thousandths digit is 7, so the hundredths digit rounds upward.

6. $\frac{2}{3}$

Divide numerator and denominator by 21.

7. $\frac{29}{36}$

The least common denominator is 36.

$$\frac{5}{12} = \frac{15}{36}$$
$$\frac{7}{18} = \frac{14}{36}$$
$$\frac{15}{36} + \frac{14}{36} = \frac{29}{36}$$

8. $\frac{2}{3}$

$$\frac{3}{5} \times \frac{10}{9} = \frac{30}{45} = \frac{2}{3}$$

9. 3.2 percent

Move the decimal two places right.

10. $396

The remaining percentage is:

$$100\% - 17.5\% = 82.5\%$$
$$480(0.825) = 396$$

11. $4:7$

Divide both terms by 7.

12. $13.50 per item

$$94.50 \div 7 = 13.50$$

13. $x = 36$

$$9 \times 56 = 14x$$
$$504 = 14x$$
$$x = 36$$

14. 165 minutes

$$2.75 \times 60 = 165$$

15. 27.3 feet

$$4.2 \times 6.5 = 27.3$$

16. $13x + 8$

Combine variable terms and constants.

17. $-3y + 12$

$$-4(2y - 3) + 5y = -8y + 12 + 5y = -3y + 12$$

18. 20

$$3(-3)^2 + 2(-3) - 1 = 27 - 6 - 1 = 20$$

19. $x = 6$

$$6x - 5 = 31$$
$$6x = 36$$
$$x = 6$$

20. $x = 7$

$$3x + 17 = 7x - 11$$
$$28 = 4x$$
$$x = 7$$

21. $x = 4$

Multiply every term by 12:

$$2x + 3 = 11$$
$$2x = 8$$
$$x = 4$$

22. -2

$$m = \frac{-7-5}{4-(-2)} = \frac{-12}{6} = -2$$

23. Mean 8, median 7.5

Mean:

$$\frac{4+6+6+9+10+13}{6} = \frac{48}{6} = 8$$

Median:

$$\frac{6+9}{2} = 7.5$$

24. Seven visits

$$22 + 14v = 120$$
$$14v = 98$$
$$v = 7$$

25. Length 12 feet, perimeter 40 feet

$$96 \div 8 = 12$$

$$P = 2(12) + 2(8) = 40$$

26. x^6

$$\frac{x^{11}}{x^5} = x^6$$

27. 4.9×10^{-6}

28. $7\sqrt{3}$

$$147 = 49 \times 3$$

$$\sqrt{147} = 7\sqrt{3}$$

29. $(x - 7)(x + 5)$

The numbers multiply to -35 and add to -2.

30. $x = 3$ or $x = -4$

$$x^2 + x - 12 = (x + 4)(x - 3)$$

Set each factor equal to zero.

Domain Review

Questions 1 through 5: Number sense

Questions 6 through 10: Fractions, decimals, percentages

Questions 11 through 15: Ratios and units

Questions 16 through 21: Algebra

Questions 22 through 25: Graphs, data, applications

Questions 26 through 30: Higher mathematics

For each miss, record the Error Pattern and next repair.

Appendix E: Practice Assessment Two

Complete this assessment near the end of the 30-day plan.

Use realistic test conditions. Questions 26 through 30 are Higher-Math questions and are optional unless your verified route requires them.

Questions

1. Evaluate:

$$8,015 - 4,978$$

2. Evaluate:

$$24 \div 3 \times 2 + 5$$

3. Evaluate:

$$-6 - (-11) + 2$$

4. Which is smaller, 0.407 or 0.47?
5. Estimate 61.2×8.1.
6. Simplify:

$$\frac{54}{72}$$

7. Subtract:

$$\frac{7}{8} - \frac{5}{12}$$

8. Multiply:

$$1\frac{1}{2} \times 2\frac{2}{3}$$

9. Convert 68 percent to a decimal and a simplified fraction.
10. A price increases from \$320 to \$368. Find the percent increase.
11. A vehicle travels 414 miles on 13.8 gallons. Find miles per gallon.
12. Five notebooks cost \$18.75. Find the cost of eight notebooks at the same unit price.
13. Solve:

$$\frac{7}{9} = \frac{42}{x}$$

14. Convert 7,920 feet to miles.
15. A model uses a scale of 1 centimeter to 2.4 meters. A length on the model is 8.5 centimeters. Find the actual length.
16. Simplify:

$$6(2x + 1) - 4(x - 3)$$

17. Evaluate $2b^2 - 5b + 4$ when $b = 3$.
18. Solve:

$$5(x - 4) = 3x + 10$$

19. Solve:

$$0.3x + 1.8 = 7.2$$

20. Solve:

$$\frac{2x-1}{5} = 3$$

21. Solve:

$$4 - 3x > 19$$

22. Find $f(-3)$ if:

$$f(x) = x^2 - 2x - 4$$

23. Find the slope and y intercept of:

$$y = \frac{3}{2}x - 7$$

24. The data are 5, 6, 8, 8, 10, 35. Find the mean and median. Which better represents a typical value?
25. Plan A costs $65 plus $4 per visit. Plan B costs $25 plus $8 per visit. At how many visits are the costs equal?
26. Simplify:

$$\left(3x^2y\right)^2$$

27. Write 6.2×10^5 in standard notation.
28. Multiply:

$$(x - 5)(x + 2)$$

29. Factor:

$$2x^2 + 10x$$

30. Solve:

$$2x^2 - 8 = 0$$

Answers and Explanations

1. $3,037$

Estimate: $8,000 - 5,000 \approx 3,000$.

2. 21

Multiplication and division proceed left to right:

$$24 \div 3 = 8$$
$$8 \times 2 = 16$$
$$16 + 5 = 21$$

3. 7

$$-6 - (-11) + 2 = -6 + 11 + 2 = 7$$

4. 0.407

Compare 0.407 with 0.470.

5. About 480

Round to $60 \times 8 = 480$. The exact answer is not required.

6. $\frac{3}{4}$

Divide numerator and denominator by 18.

7. $\frac{11}{24}$

$$\frac{7}{8} = \frac{21}{24}$$
$$\frac{5}{12} = \frac{10}{24}$$
$$\frac{21}{24} - \frac{10}{24} = \frac{11}{24}$$

8. 4

$$1\frac{1}{2} = \frac{3}{2}$$
$$2\frac{2}{3} = \frac{8}{3}$$
$$\frac{3}{2} \times \frac{8}{3} = 4$$

9. 0.68 and $\frac{17}{25}$

$$68\% = \frac{68}{100} = \frac{17}{25}$$

10. 15 percent

Increase:

$$368 - 320 = 48$$

Percent increase:

$$\frac{48}{320} \times 100\% = 15\%$$

11. 30 miles per gallon

$$414 \div 13.8 = 30$$

12. $30

Unit price:

$$18.75 \div 5 = 3.75$$

Eight notebooks:

$$3.75 \times 8 = 30$$

13. $x = 54$

$$7x = 9(42)$$
$$7x = 378$$
$$x = 54$$

14. 1.5 miles

$$7,920 \div 5,280 = 1.5$$

15. 20.4 meters

$$8.5 \times 2.4 = 20.4$$

16. $8x + 18$

$$12x + 6 - 4x + 12 = 8x + 18$$

17. 7

$$2(3)^2 - 5(3) + 4 = 18 - 15 + 4 = 7$$

18. $x = 15$

$$5x - 20 = 3x + 10$$
$$2x = 30$$
$$x = 15$$

19. $x = 18$

$$0.3x = 5.4$$
$$x = 18$$

20. $x = 8$

$$2x - 1 = 15$$
$$2x = 16$$
$$x = 8$$

21. $x <- 5$

$$4 - 3x > 19$$
$$- 3x > 15$$

Divide by -3 and reverse the inequality:

$$x <- 5$$

22. 11

$$f(-3) = (-3)^2 - 2(-3) - 4$$
$$= 9 + 6 - 4 = 11$$

23. Slope $\frac{3}{2}$, y intercept -7

24. Mean 12, median 8. The median better represents a typical value because 35 an outlier that pulls the mean upward.

$$\frac{5+6+8+8+10+35}{6} = \frac{72}{6} = 12$$

25. 10 visits

$$65 + 4v = 25 + 8v$$
$$40 = 4v$$
$$v = 10$$

Check:

$$65 + 4(10) = 105$$
$$25 + 8(10) = 105$$

26. $9x^4y^2$
27. 620,000
28. $x^2 - 3x - 10$
29. $2x(x + 5)$
30. $x = 2$ or $x =- 2$

$$2x^2 - 8 = 0$$
$$2x^2 = 8$$
$$x^2 = 4$$
$$x =\pm 2$$

Final Assessment Review

For every missed question, write:

Skill tested:

First wrong step:

Error Pattern:

Prevention rule:

Action before test day:

Select no more than three High-Value Skills for final repair.

Appendix F: Error Pattern Log

The Error Pattern Log turns wrong answers into instructions.

Copy the table into your notebook or print several copies.

Date	Problem	First Wrong Step	Error Pattern	Preventi on Rule	Retest Date	Result

Error Pattern Categories

Use a category that describes the actual failure.

Sign error

A negative sign was lost, added, or handled incorrectly.

Operation choice

The mathematics performed did not match the relationship.

Fraction rule

Denominators, reciprocals, simplification, or mixed numbers were handled incorrectly.

Decimal placement

Place value was lost during conversion or calculation.

Copying

A number, exponent, unit, or operation changed between lines.

Vocabulary

The instruction or algebraic language was misunderstood.

Formula selection

The wrong relationship was chosen.

Skipped step

A transformation happened mentally and created an error.

Rushed reading

The requested quantity or restriction was missed.

Answer entry
The scratch work was correct but the response was entered incorrectly.

Unknown concept

You did not know how to begin.

Strong Prevention Rules

Weak: Be careful with signs.
Strong: Circle every negative sign before the first calculation.
Weak: Read better.
Strong: Underline the final question and write the requested unit.
Weak: Practice fractions.
Strong: Do not add until the denominators match.
Weak: Check my work.
Strong: Substitute the answer into the original equation.
Weak: Stop making calculator errors.
Strong: Write the expression on paper before entering it.

Sample Log Entry

Date	Problem	First Wrong Step	Error Pattern	Prevention Rule	Retest Date	Result
Day 9	15 percent discount on \$80	Reported \$12 instead of \$68	Rushed reading	Circle change or final amount	Day 11	Correct
Day 15	$7 - 2x =$	Divided by 2 without carrying negative	Sign error	Rewrite $-2x = 6$ on its own line	Day 16	Correct
Day 20	Slope from two points	Reversed subtraction in denominator only	Formula use	Keep point order consistent in top and bottom	Day 22	Correct

Weekly Pattern Review

At the end of each week, count the categories.

Most common pattern:
Prevention rule used:
Evidence that the rule worked:
Pattern that still repeats:
Help needed:
Next retest date:

A repeated error is not evidence that you cannot learn the topic. It is evidence that the current prevention rule has not yet changed the behavior.

Appendix G: Formula and Vocabulary Sheet

Use this sheet for review. Confirm which formulas are provided or expected on your exact Placement Test.

Number and Percent Relationships

$$part = percent\ as\ a\ decimal \times whole$$

$$percent\ change = \frac{new - original}{original} \times 100\%, \quad where\ the\ original\ value\ is\ not\ zero$$

Percent increase multiplier:

$$1 + rate\ as\ a\ decimal$$

Percent decrease multiplier:

$$1 - rate\ as\ a\ decimal$$

Rate and Proportion

$$unit\ rate = \frac{quantity}{one\ unit}$$

$$\frac{a}{b} = \frac{c}{d}, \quad b \neq 0,\ d \neq 0$$

For a true proportion with nonzero denominators:

$$ad = bc$$

Geometry

Rectangle perimeter:

$$P = 2l + 2w$$

Rectangle area:

$$A = lw$$

Triangle area:

$$A = \frac{1}{2}bh$$

Circle circumference:

$$C = 2\pi r$$

Circle area:

$$A = \pi r^2$$

Right triangle:

$$a^2 + b^2 = c^2$$

Distance Rate Time

$$d = rt$$
$$r = \frac{d}{t}, \quad t \neq 0$$
$$t = \frac{d}{r}, \quad r \neq 0$$

Linear Relationships

Slope:

$$m = \frac{y_2 - y_1}{x_2 - x_1}, \quad x_2 \neq x_1$$

Slope intercept form:

$$y = mx + b$$

Proportional relationship:

$$y = kx$$

Statistics

Mean:

$$mean = \frac{sum\ of\ values}{number\ of\ values}$$

Range:

$$maximum - minimum$$

Exponents

$$x^a x^b = x^{a+b}$$

For $x \neq 0$:

$$\frac{x^a}{x^b} = x^{a-b}$$
$$\left(x^a\right)^b = x^{ab}$$
$$x^0 = 1$$
$$x^{-n} = \frac{1}{x^n}$$

Quadratic Formula

For $ax^2 + bx + c = 0$, where $a \neq 0$:

$$x = (-b \pm \sqrt{(b^2 - 4ac)}) / 2a$$

Use only if relevant to your verified route.

Canonical Vocabulary

Placement Test: An assessment or guided process used by an institution to help determine an appropriate starting course or level of support.

Placement Policy: The institution's rules for using scores, transcripts, prior courses, retests, exemptions, and other measures.

Target Course: The first mathematics course you actually need for your program or academic route.

Placement Gap: The difference between your current demonstrated skill and the skill required for the Target Course.

Five-Rung Placement Ladder: Clarify, Diagnose, Rebuild, Rehearse, and Decide.

No-Shame Diagnostic: The initial skills assessment classified as Ready, Recall, or Rebuild rather than pass or fail.

Ready Skill: A skill you can perform accurately without immediate review.

Recall Skill: A previously learned skill that returns after a prompt or short refresher.

Rebuild Skill: A skill requiring direct instruction and repeated practice.

High-Value Skill: A skill that appears on the relevant assessment or supports several later skills.

Error Pattern: A repeated type of mistake.

Error Pattern Log: The worksheet used to record, classify, correct, and prevent repeated errors.

Mixed Practice: Practice that combines several topics so you must choose the method.

Calm Test Routine: Settle, Read, Mark, Represent, Solve, Check, and Release.

Final Readiness Check: The last review of skills, logistics, policies, and test day preparation.

Appendix H: Placement Research Worksheets

Placement Policy Research Sheet

Institution:

Campus:

Program:

Enrollment term:

Placement process or test:

Sections required:

Placement affects:

Admission:

Course registration:

Support requirement:

Other:

Current score or placement requirement:

Does the institution set its own score?

Multiple measures used:

Prior college coursework accepted:

High school transcript considered:

Other test scores accepted:

Guided placement available:

Exemption rules:

Retest allowed:

Number of attempts:

Waiting period:

Required learning before retest:

Calculator rule:

Scratch paper rule:

Remote testing requirements:

Accessibility contact:

Official testing contact:

Official page title:

Date verified:

Questions still unanswered:

Target Course Map

Program:

Catalog year:
First required mathematics course:
Course number:
Alternative accepted courses:
Prerequisites:
Placement needed:
Minimum grade:
Course that follows:
Program deadline affected:
Opening week skills:
Tutoring available:
Corequisite support available:
Compressed or full term:
Transfer considerations confirmed:
Adviser name or office:
Date verified:

Adviser Question List

Use the questions relevant to your situation.

1. Which mathematics course is the first required course for my program?
2. Does my placement result affect admission or only course registration?
3. Does the college use prior coursework, grades, or other measures?
4. Is a supported or corequisite version of the Target Course available?
5. May I retest?
6. What must I complete before retesting?
7. How long is the result valid?
8. Will the assigned foundation course count toward my credential?
9. Could this course affect financial aid or program timing?
10. What skills should I review before the Target Course begins?
11. What tutoring or mathematics lab support is available?
12. Who can review a placement that appears inconsistent with my record?

Contact Script

"I am entering the __________ program for the __________ term. The first mathematics course listed in my program is __________. I am trying to confirm the current placement process, whether prior coursework or other measures a considered, and what retest or support options are available. Could you direct me the current written policy or the office responsible for this decision?"

Date Stamp Rule

Write a date beside every live policy fact.

Recheck:

two weeks before testing

before paying a test fee

before a retest

before registration

before publication of any future edition of this book

Appendix I: Official Resource Directory

Use current official resources first. Commercial resources can add instruction and practice, but they should not replace the institution's current Placement Policy.

ACCUPLACER

College Board provides official descriptions of the Arithmetic; Quantitative Reasoning, Algebra, and Statistics; and Advanced Algebra and Functions tests (College Board, n.d.-k).

College Board provides free practice tests, downloadable sample questions, and learning resources (College Board, n.d.-c). Its test-day page links to test-taking tutorials (College Board, n.d.-f), and its calculator guidance explains the onscreen-calculator and handheld-calculator rules (College Board, n.d.-a).

The institution sets local placement scores and policies. ACCUPLACER does not impose one universal course placement score (College Board, n.d.-i).

ALEKS PPL

McGraw Hill describes ALEKS PPL as an adaptive, open-response assessment (McGraw Hill, n.d.-b) and describes its current personalized preparation and learning component on the higher-education product page (McGraw Hill, n.d.-a).

Institutional configuration may differ. Confirm attempts, waiting periods, learning requirements, proctoring, score bands, and calculator use with your college.

TSIA2

College Board's TSIA2 student brochure explains the assessment's college-readiness purpose and test features (College Board, 2020).

College Board hosts the required preassessment activity, free TSIA2 practice tests, and learning resources (College Board, n.d.-d).

Check current Texas exemption and readiness rules through official state and institutional sources.

PERT

Florida's Department of Education provides current information about demonstrating college readiness and describes PERT as an untimed computer-adaptive assessment that is not pass or fail (Florida Department Education, n.d.).

Confirm whether PERT is required for your specific institution and stude status.

College and State Resources

Your highest priority local sources are:

testing services
assessment center
academic advising
mathematics department
current college catalog
current program map
registrar
financial aid office for aid questions
disability or accessibility services
state higher education agency where applicable

Publication Date Recheck List

Before publishing or updating this book, recheck:

1. ACCUPLACER test names and content descriptions
2. ACCUPLACER timing language
3. ACCUPLACER calculator policy
4. ACCUPLACER score and retest language
5. ALEKS PPL question count and product features
6. ALEKS PPL attempt and access language
7. TSIA2 student information and practice pages
8. Texas readiness and exemption rules
9. Florida PERT format and policy
10. Corequisite and multiple measures descriptions
11. Every state or college example
12. Every trademark statement
13. Every live URL in the references
14. All publication years and access dates

15. The claim that no universal score applies

Do not present a local cut score as national.

Do not copy a college policy into a future edition without a date and recheck.

Appendix J: Final Readiness Pages

Seven-Day Checklist

Seven days before:

Practice Assessment Two completed.
Three High-Value Skills selected.
Repairs scheduled.
Six days before:
Test date and time confirmed.
Location or login confirmed.
Identification confirmed.
Calculator and scratch paper rules confirmed.
Accessibility arrangements confirmed.
Five days before:
Repair one completed.
Delayed retest scheduled.
Four days before:
Repair two completed.
Error Pattern Log reviewed.
Three days before:
Repair three completed.
Short Mixed Practice completed.
Two days before:
No new domain added.
Formula and vocabulary review completed.
Materials and travel prepared.
One day before:
Light review only.
Calm Test Routine rehearsed.
Stopping time chosen.

Calm Test Routine Card

Settle

Feet on floor. Slow exhale.

Read

Read the full question.

Mark

Requested quantity, signs, units, restrictions.

Represent

Equation, diagram, table, estimate, or first step.

Solve

One line at a time.

Check

Sign, size, unit, context.

Release

Submit or move on according to the rules.

After the Result

Score or placement:
- Target Course:
- Assigned course:
- Domain information:
- Retest allowed:
- Waiting period:
- Learning required:
- Multiple measures available:
- Corequisite support:
- Adviser contact:
- Registration deadline:
- Next action:

College Mathematics Continuation Plan

First class date:

Course:
Three skills to maintain:
Two skills to rebuild:
Weekly practice times:
Tutoring location or link:
Instructor office hours:
Study partner or group:
First progress review date:
Most important prevention rule:

Final Statement

My placement is the course where I begin, not the limit of where I can go.

References

Barroso, C., Ganley, C. M., McGraw, A. L., Geer, E. A., Hart, S. A., & Daucourt, M. C. (2021). A meta-analysis of the relation between math anxiety and math achievement. *Psychological Bulletin, 147*(2), 134-168. https://doi.org/10.1037/bul0000307

College Board. (2020). *Texas Success Initiative Assessment 2.0: Student informational brochure*. https://accuplacer.collegeboard.org/accuplacer/pdf/tsia2-student-informational-brochure.pdf

College Board. (n.d.-a). *Can I use calculators on the ACCUPLACER math tests?* ACCUPLACER. Retrieved July 12, 2026, from https://accuplacer.collegeboard.org/help-center/can-i-use-calculators-accuplacer-math-tests

College Board. (n.d.-b). *Develop placement policies*. ACCUPLACER. Retrieved July 12, 2026, from https://accuplacer.collegeboard.org/educators/implementing-accuplacer/develop-placement-practices

College Board. (n.d.-c). *Free ACCUPLACER practice resources*. ACCUPLACER. Retrieved July 12, 2026, from https://accuplacer.collegeboard.org/students/prepare-for-accuplacer/practice

College Board. (n.d.-d). *Free TSIA2 PAA and practice resources*. ACCUPLACER. Retrieved July 12, 2026, from https://accuplacer.collegeboard.org/students/free-tsia2-pre-assessment-and-practice-resources

College Board. (n.d.-e). *Get my scores*. ACCUPLACER. Retrieved July 12, 2026, from https://accuplacer.collegeboard.org/students/scores/getting-scores

College Board. (n.d.-f). *Get ready for test day*. ACCUPLACER. Retrieved July 12, 2026, from https://accuplacer.collegeboard.org/students/before-test-day/test-day

College Board. (n.d.-g). *Get to know ACCUPLACER*. ACCUPLACER. Retrieved July 12, 2026, from https://accuplacer.collegeboard.org/about/get-to-know-accuplacer

College Board. (n.d.-h). *How multiple factors improve placement decisions*. ACCUPLACER. Retrieved July 12, 2026, from https://accuplacer.collegeboard.org/educators/make-holistic-placement-decisions

College Board. (n.d.-i). *Set course placement scores*. ACCUPLACER. Retrieved July 12, 2026, from https://accuplacer.collegeboard.org/educators/implementing-accuplacer/set-cc se-placement-scores

College Board. (n.d.-j). *Understand my scores*. ACCUPLACER. Retrieved July 1 2026, from https://accuplacer.collegeboard.org/students/scores/understanding-scores

College Board. (n.d.-k). *What's on the tests*. ACCUPLACER. Retrieved July 12, 2026, from https://accuplacer.collegeboard.org/students/prepare-for-accuplacer/whats-on-t sts

Community College Research Center. (2021, July). *Developmental education*. Teachers College, Columbia University. https://ccrc.tc.columbia.edu/publications/developmental-education.html

Florida Department of Education. (n.d.). *Demonstration of college readiness (formerly common placement testing/PERT)*. Retrieved July 12, 2026, from https://www.fldoe.org/schools/higher-ed/fl-college-system/academics/common- placement-testing.stml

Kopko, E. M., & Daniels Sarica, H. (2023, October). *The long-term effects of multiple measures assessment at SUNY community colleges*. Community College Research Center, Teachers College, Columbia University. https://ccrc.tc.columbia.edu/publications/long-term-effects-multiple-measures-a ssessment.html

Lyle, K. B., Bego, C. R., Ralston, P. A. S., & Immekus, J. C. (2022). Spaced retrieval practice imposes desirable difficulty in calculus learning. *Educational Psychology Review, 34*, 1799-1812. https://doi.org/10.1007/s10648-022-09677-2

McGraw Hill. (n.d.-a). *ALEKS Placement, Preparation and Learning*. Retrieved July 12, 2026, from https://www.mheducation.com/highered/digital-products/alekssppl.html

McGraw Hill. (n.d.-b). *ALEKS PPL: A math placement test*. Retrieved July 12, 2026, from https://www.mheducation.com/prek-12/program/microsites/MKTSP-GAB02M0 /browse/ppl.html

National Center for Education Statistics. (n.d.). *Student enrollment: Age of students* [Data tool]. IPEDS Trend Generator. Retrieved July 12, 2026, from https://nces.ed.gov/ipeds/trendgenerator/app/answer/2/8

Ran, F. X. (2020, January 29). *Corequisite remediation is shown to improve gateway course outcomes. Now what?* Community College Research Center, Teachers College, Columbia University. https://ccrc.tc.columbia.edu/easyblog/future-of-corequisite-remediation.html